RESTORING THE VICTORIAN HOUSE

AND OTHER TURN-OF-THE-CENTURY STRUCTURES

RESTORING THE VICTORIAN HOUSE

AND OTHER TURN-OF-THE-CENTURY STRUCTURES

JOHN BRACKEN
WITH LINDA STONE

Chronicle Books / San Francisco

Printed in the United States of America

Library of Congress Cataloging in Publication Data
Bracken, John, 1927–
Restoring the Victorian house.
Includes index.
1. Buildings — United States — Repair and reconstruction.
2. Architecture, Victorian — United States. I. Stone, Linda.
III. Title.
TH3411.B69 690'.837'0288 81-2452
ISBN 0-87701-222-9 (pbk.) AACR2

Design: Goodchild Jacobsen
Editing: Philip C. Johnson
Composition: Type by Design, Fairfax, California
Photography: Chalmers and Dean Luckhardt
Cover photography: Curt Bruce
Drawings: Linda Stone

Chronicle Books
870 Market Street
San Francisco, California 94102

CONTENTS

RESTORING THE VICTORIAN HOUSE

AND OTHER TURN-OF-THE-CENTURY STRUCTURES

PART ONE

THE VICTORIAN HOUSE

It is strange that in the first years of Queen Victoria's reign (1837-1901) the dominant architecture was certainly not Victorian; it was an extension of the Regency style (1811-1830), with classical lines and stucco facades. The stucco was complemented by decorative iron balconies, and always there were bay or bow windows. It was a classical, square look, with blocks of stucco or yellow brick and slate roofs. But in London some changes in the neo-Regency style were taking place as houses became narrower in size because of lack of ground space. Thomas Cubitt and James Burton were the most prominent builders for the middle class. Their homes were fashioned after the eighteenth century Georgian manner — good basic design — but the buildings were somewhat ornate and lacked the charm and lightness of the earlier periods. This style prevailed until 1840, when Gothic Revivalism started to creep in and became the prevailing style of architecture until 1860, when the patterns became Victorian Gothic with the use of greater ornamentation.

The crucial moment in American architecture started about 1800, with the arrival of professional architects from abroad. Before this time, American architecture had been 50 to 100 years behind the English style.

After the American Revolution (1775-1783), the basic concept of Colonial (Georgian) architecture began to change somewhat. English building books had been the basic source for American buildings, with the Georgian style (1702-1830) popular. When the Capitol and White House were conceived, it was necessary to hire foreign architects for the work because there were no professional architects in America. Washington and Jefferson created the Federal style, but by 1820 the Georgian and Federal Colonial styles had almost disappeared (not to be revived until the end of the century), and the Greek Revival movement started to prevail. For 40 years, until the advent of the Civil War, it was the style for most American buildings of any stature. Greek Revival predominated in Philadelphia; homes there were not as large or richly decorated as those in England, but classical ornamentation was used frequently.

A London-trained architect, Latrobe, was responsible for the first Gothic Revival house in America. The predominate features of this style were arched windows and doors; a vertical exterior form, with a steep roof and chimneys; and picturesque detailing. Decorative details were added with abandon and with no eye to authenticity. If it pleased the eye, it was acceptable to the Americans.

During the Civil War (1861-1865), many fortunes had been amassed, and the newly rich wanted to show their grandness. After the war, major structural and mechanical developments made possible great advances in building. A new architecture evolved that was akin to Italianate but inspired by the contemporary European styles of French Second Empire. The post-Civil War period was an age of extensive mass production and stock designs. Once designed, a fine townhouse could then be reproduced in rows by almost anyone handy with a hammer. Thus, long rows of brownstones rose quickly.

Toward the end of the 1860s architects entered a phase of romanticism that was expressed in Gothic Victorian style. Not since the Gothic Revival and the Italianate styles of the 1840s and 1850s had Americans plunged into such a wave of romanticism. By the 1880s the style was flamboyant, fluffy, and decorative beyond description, unlike the somber style of the 1850s. Mansard roofs were crowned with ornate iron railings. A form of the European Queen Anne style (1777-1781), with its bowed windows and round curves, became immensely popular. New houses combined all kinds of materials, with porches, balconies, towers, and wrought iron used with abandon. Now even the middle classes could well afford the $1700 necessary for a Victorian house, whether is was Gothic, Queen Anne, or Italianate. These houses were bastardized versions of the main styles; they were of smaller proportions, but with no less ornamentation and embellishments.

[CHAPTER 1]

VICTORIAN STYLES: PAST AND PRESENT

The word "Victorian" conjures up all sorts of images in people's minds. To some people the word denotes a period of staidness, of proper manners and strict morals. Others are reminded of gingerbread, referring to the excessive ornamentation of the Victorian building. But to many people, "Victorian" means the period from 1837 to 1901, when Queen Victoria reigned over Great Britain, Ireland, and eventually, India. Arts and invention thrived under her influence, exemplified in the magnificent Crystal Palace exhibit in 1853. In this book we use the word Victorian in a broad sense. In America, Victorian architecture denoted anything from gingerbread ornamentation to Neoclassical decorations. It was a potpourri of various architectural elements and styles.

"Victorian" also denoted opulence. The monetary riches from gold and silver, cattle and wheat were translated into the physical sense as homes as spacious and extraordinary as any parcel of land would hold. By the 1800s, the new wealthy pioneers in America went all out for more, more, and still more decoration and dramatic eye appeal. Building facades were laced with spindles and brackets, stained glass, columns, and porticoes. Inside the structures were conglomerations of fringed lampshades, patterned rugs, statuary, paneling, rich upholstery, and ornately carved furniture. The blatant use of such extreme ornamentation was a way of exhibiting the enthusiasm and the excitement of the times. Italianate, Stick, Queen Anne, Classic (and Neoclassic), and other styles became the rage through the years.

[VICTORIAN CONSTRUCTION]

The year 1839 heralded machine-made nails, standardized lumber, and a new style of construction known

Handsomely appointed with columns and frescoes, this Italianate Victorian is a beautiful monument to the past and well restored.

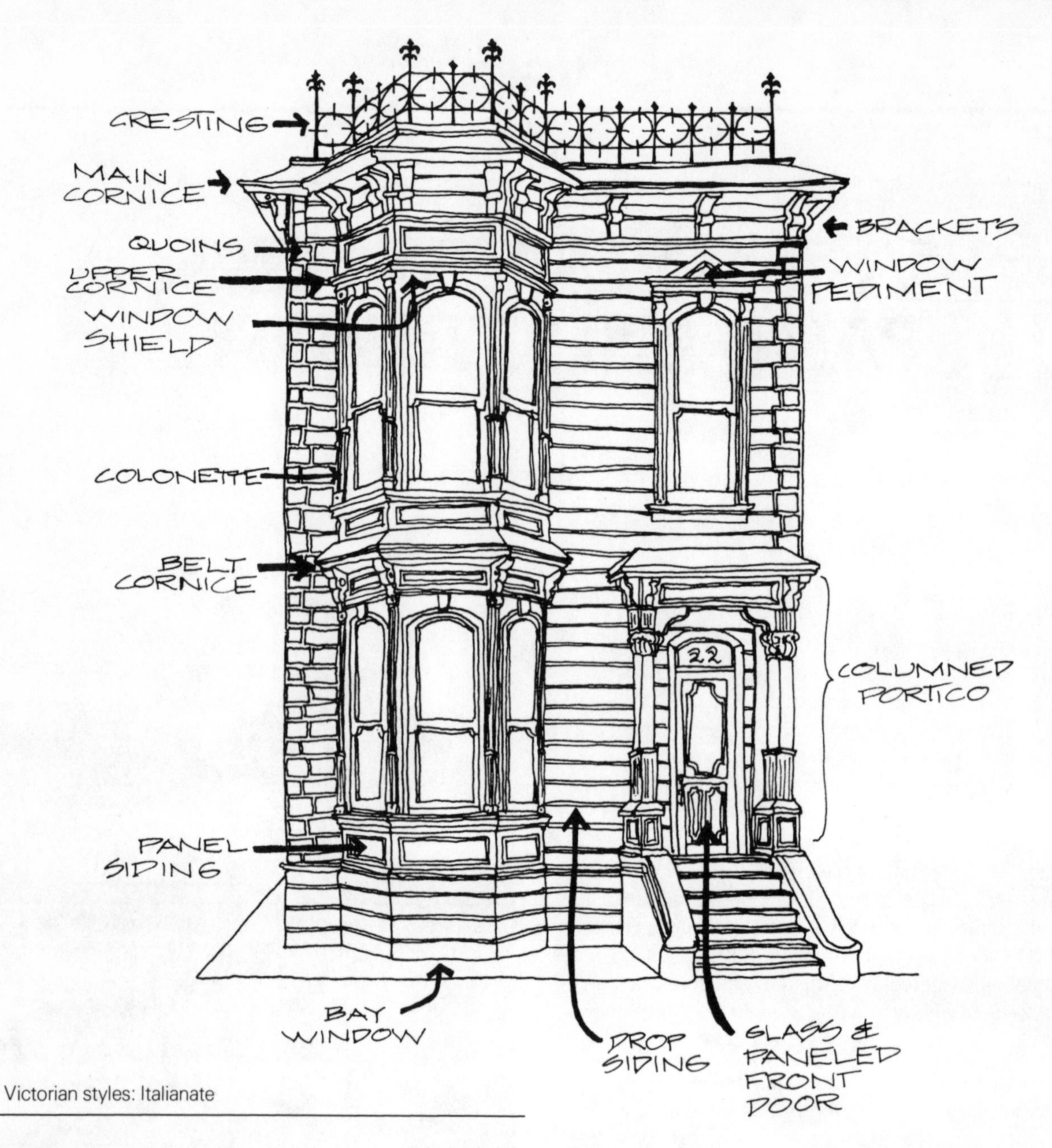

Victorian styles: Italianate

as balloon framing. No more were massive posts necessary to anchor structures to the ground; no cross beams or intricate joinery of *mortise* and *tenon* connections had to be considered. Construction of the Victorian house, previously done by master builders, now was within the domain of almost any person capable of handling a hammer and nails. A variation, called platform framing, where one story was built at a time, with each floor acting as a platform for the next, made it possible for only two people to complete a house.

Balloon framing uses lightweight lumber and wire nails instead of intricate joints. The vertical members — the studs — extend from the foundation to the roof in one piece; horizontal members — joists — are nailed to the studs. The siding or diagonal sheathing gives a structure its rigidity. Generally used for two-story houses, the balloon framing is easy to build. All you need is a hammer, nails, saw, and tape.

Also about 1839 mass-produced, elaborate building ornaments were being manufactured and advertised in house parts catalogs. This enabled people to select personalized ornamentation, whether a cornice, frieze, or other decoration.

[ITALIANATE STYLE]

Between the mid-1860s and 1880s, the stately and tall, two-story, very vertical structure with slanted bays and a raised porch made its appearance. The Italianate home's heritage was the stone structures of fifteenth and sixteenth century Italy. The stone was translated into redwood, and the square cornices were decorative translations of the original masonry. The Victorian Italianate style usually had wood Corinthian columns on the porch rather than the original carved marble. The facade also had brackets, panels, and keystones. Coved ceilings

Windows appear more like sculptural reliefs in this Italianate house.

This Italianate Victorian is beautifully ornamented—note the handsome windows with Baroque influence.

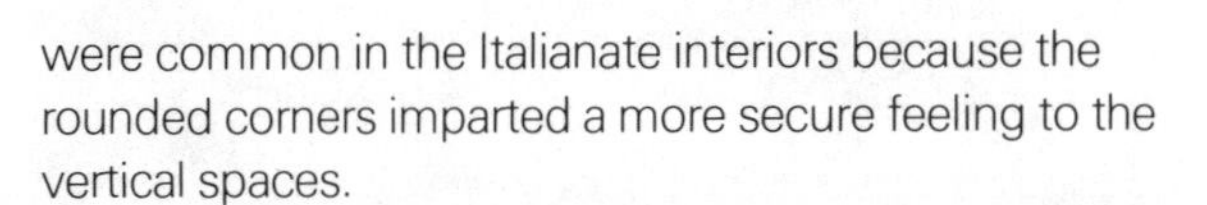

were common in the Italianate interiors because the rounded corners imparted a more secure feeling to the vertical spaces.

The gracefully arched windows on the Italianate house were almost always sculptural, and the long and narrow openings assumed a three-dimensional quality. There were columns at either side of the openings, a decorative shield above, and a segmented hood or cornice to top the window off. It was really a work of immense decorativeness. The porch had carved newel posts and an overhead portico. Inside the house there was a long hall, with the parlor to the left, usually followed by two other parlors, one to the dining room. Within the rooms were arched passageways, and sliding wooden doors were used for privacy as well as for fuel conservation. Wainscoting prevailed. The coved ceilings were 12 feet or higher.

The three-dimensional quality of the windows and the graceful proportions of this structure are all part of the Italianate style.

[SAN FRANCISCO STICK STYLE]

Victorian styles: San Francisco stick

The usual Stick Victorian was a structure of straight lines and right angles, popular from the 1870s to the 1890s. Narrow boards were used on the outside of the building for reinforcement. The porch usually had diagonal braces installed parallel to the facade. The roof extended over the house front, so the gable ends formed a separate plane. Sunray and starburst decoration was popular. Instead of the slanted bay window of the Italianate style, the Stick used a rectangular bay window, which was easier to construct.

By the 1870s wood was no longer being used to imitate stone. The Stick style was a product of scrollwork and jigsaw. A variation of this design was the so-called Eastlake style, which had excessive floral and foliage ornamentation and voluptuous columns and spindles. Charles Eastlake always referred to this style as a bastardization of his work, which it was. He started the three-dimensional, decorative wood period, turning out knobs and newels by using chisels and gouges.

The San Francisco Stick style, above, as its name suggests, emphasizes straight lines and right angles; houses conform to narrow lots and still appear appealing in proportion and design. Set on a corner this Stick style Victorian retains the straight and narrow look. Not as richly ornamented as other Victorians, the building is still handsome in all aspects.

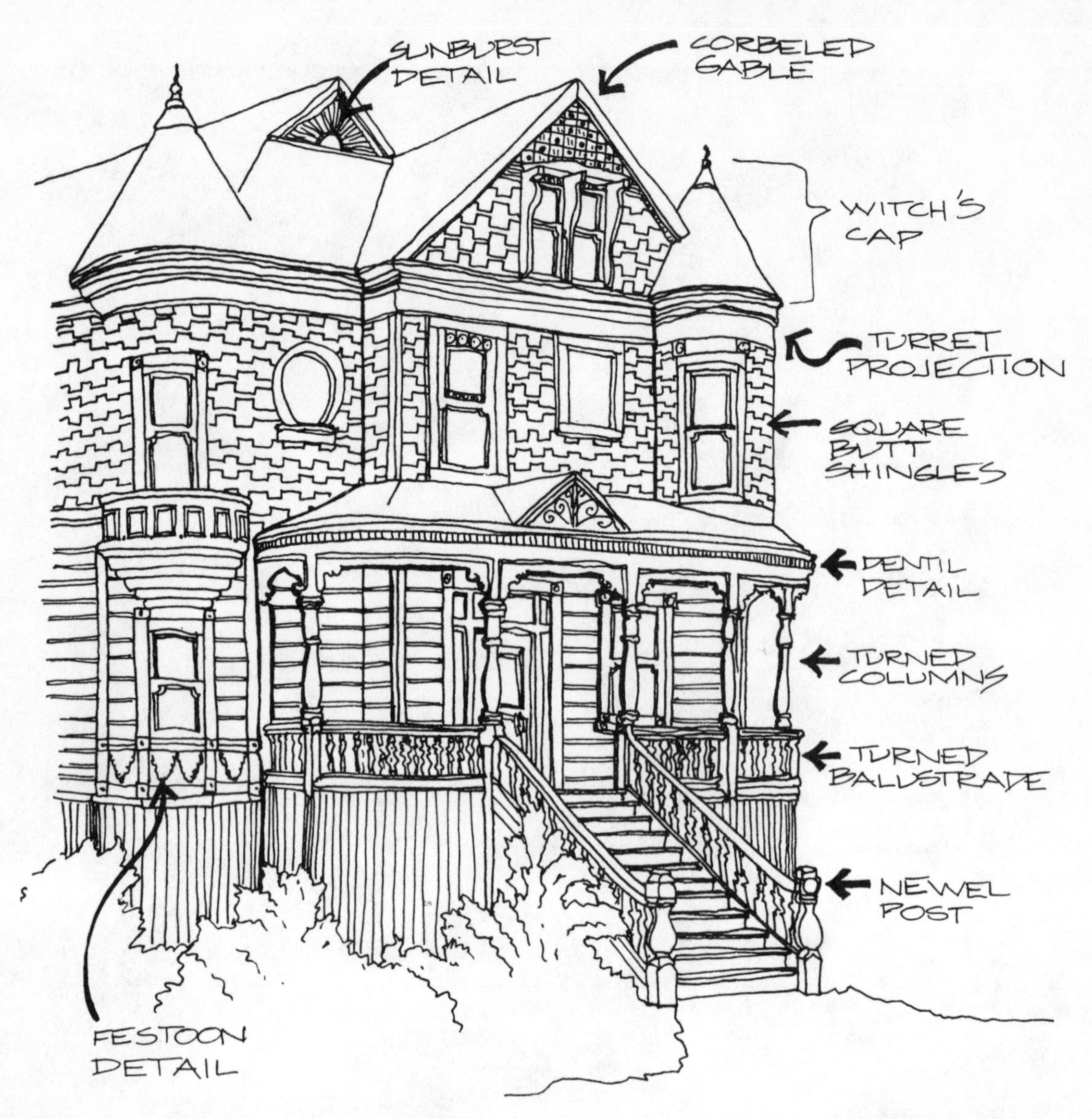

[QUEEN ANNE STYLE]

The American form of the English Queen Anne style, popular from 1883 to 1890, was more horizontal than vertical in concept and was a mixture of more volume and texture than, say, the Stick or Italianate styles. Corner towers had witches' caps and appliqued sunbursts. Balconies with turned balustrades loomed over front porches trimmed with trellage. Generally there was no single roof line, just an arrangement of varying shapes that somehow managed to look handsome. The Queen Anne house was not severe or sterile; it was a lovely lady in full regalia, with billowing blouses and artfully patterned materials all thrown together. The Queen Anne house usually had a grand hall as the core, with other rooms radiating from it.

The Queen Anne cottage, smaller and more demure looking, was usually of one story, with oversized gables. The gable was always ornamented, often times with scallops. There was a front bay window, along with cutaway corners, and a recessed porch.

Victorian styles: Queen Anne

A typical Queen Anne Structure — circular turrets and arched entryway.

Left: This Queen Anne cottage has oversized gables and porch arches — more confined than the classic Queen Anne but still an appealing structure.

This handsome Queen Anne house shows the typical curvilinear features and witch's cap of the style. The house looks large, round, and blatant.

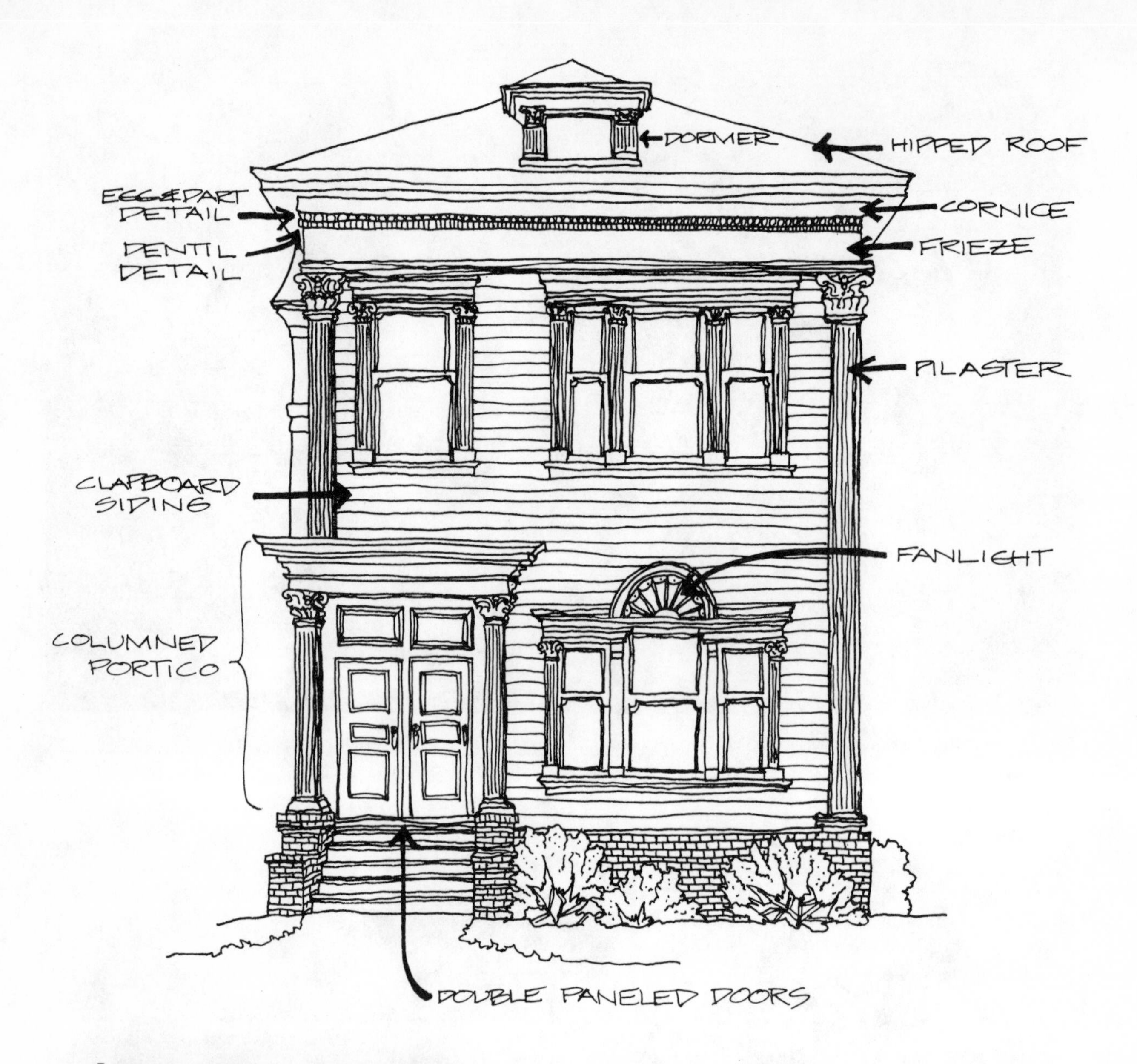

[COLONIAL REVIVAL STYLE]

By the end of the nineteenth century, the country's home-builders, somewhat sated by the excesses of the Italian-ate and Queen Anne styles, looked for a more refined style, one clean in character. America turned to its own history to find an architectural mark. The charm of the Colonial houses was carried throughout the nation as train travel improved. These Colonial homes used Roman and Greek elements: symmetrical facades, sensitive proportions, and little ornamentation. Known as Georgian and Federal, the original Colonial styles built between the Revolution and 1820 had classic columns, porticoes, and rectangular windows with small panes of glass. The 1893 Columbian Exposition in Chicago emphasized the Neo-classical fad. Greek and Roman architecture was the theme of the fair, and for years after, all kinds of structures, from public buildings to residences, used classical features such as capitals, pediments, and colonnades. (The Exposition also marked the end of the Colonial Revival period.)

Although the Colonial Revival style had a lot going for it, it did not find favor in the West. The style eventually gave way to Mission Revival.

Both Queen Anne and Neoclassical styles are married in this Victorian.

Here is a good example of the Victorian structure with Neoclassic lines from the columned windows to the frescoed adornments on the face of the building.

[CLASSIC BOX]

At the turn of the nineteenth century there was a great population burst, and the Classic style became popular from 1890 to 1910. Aptly named but certainly not descriptive of its beauty, the Classic Box resulted from fashionable trends and site restrictions. The once ornate, sprawling Queen Anne style was confined in an orderly arrangement, with flush panels, flattened ornamentation, and few protruding parts. Even the ubiquitous bay windows became flattened.

Inside, the floor plan imitated the colonists' homes: Generally, the front door was to one side and led to a foyer that had a U- or L-shaped staircase. Behind the foyer was a closet, then the bathroom, and at the rear, the kitchen. On the other side of the plan, the living room, dining room, and porch were lined up.

The Classic Box style usually had a broad peak and a dormer window in the center. The eaves were generally enclosed at the cornice, and there were swag garlands and plaster ribbons. Homes were approaching the era of the craftsperson rather than clinging to the Victorian past. Window shapes varied considerably—short and broad windows, square-looking double-hung windows—and the windows were arranged fairly symmetrically in front but scattered on the sides.

The Classic Box style of architecture shows well in this building in Louisville, Kentucky. (Courtesy of Country Preservation Alliance)

(OTHER STYLES)

The styles we discussed were the most prevalent ones, but mixed with them throughout the Victorian period, especially from 1895 on, were other styles. Among them were the Gothic with steeply pitched roofs and turrets and ornate embellishments, and the Neoclassic row house, a one-story house usually (but not always) with a hipped roof and dormer windows reminiscent of Queen Anne detailing. These structures were detached houses built side by side in cities.

The Eastlake influence is seen in this Stick style Victorian—clean simple lines and handsome decoration.

This Victorian in Louisiana has a little bit of everything going for it—Gothic and Italian Renaissance decoration on the same house. (Courtesy of Country Preservation Alliance)

The high-pitched roof lines and various turrets and dormers show the Gothic influence on this Victorian

(BROWNSTONES)

In the row house concept were the brownstones. After the middle of the nineteenth century, the wooden houses of the East (mainly Brooklyn, but in other cities as well) began to give way to brick and brownstone. Strictly speaking, brownstones were houses made of brick and faced with 4 to 6 inches of thick veneer of Connecticut River Valley or New Brunswick reddish-brown sandstone, built mainly from just after the Civil War to the early years of this century. However, in the parlance of the brownstone revival movement, the term brownstone itself has come to mean any nineteenth century rowhouse, particularly of the Victorian period, even though it may have wood shingles, clapboard, or limestone facing.

Because land became scarce in cities and building sites narrow, the row house type of architecture (one house next to the other) became popular.

The Gothic Revival influence is shown in this Victorian brownstone in New York. Note the elaborate cornices, entablatures, and cast-iron railings. (Courtesy of Landmark Preservation Commission)

Floor layout (interior): single story (below) and four story

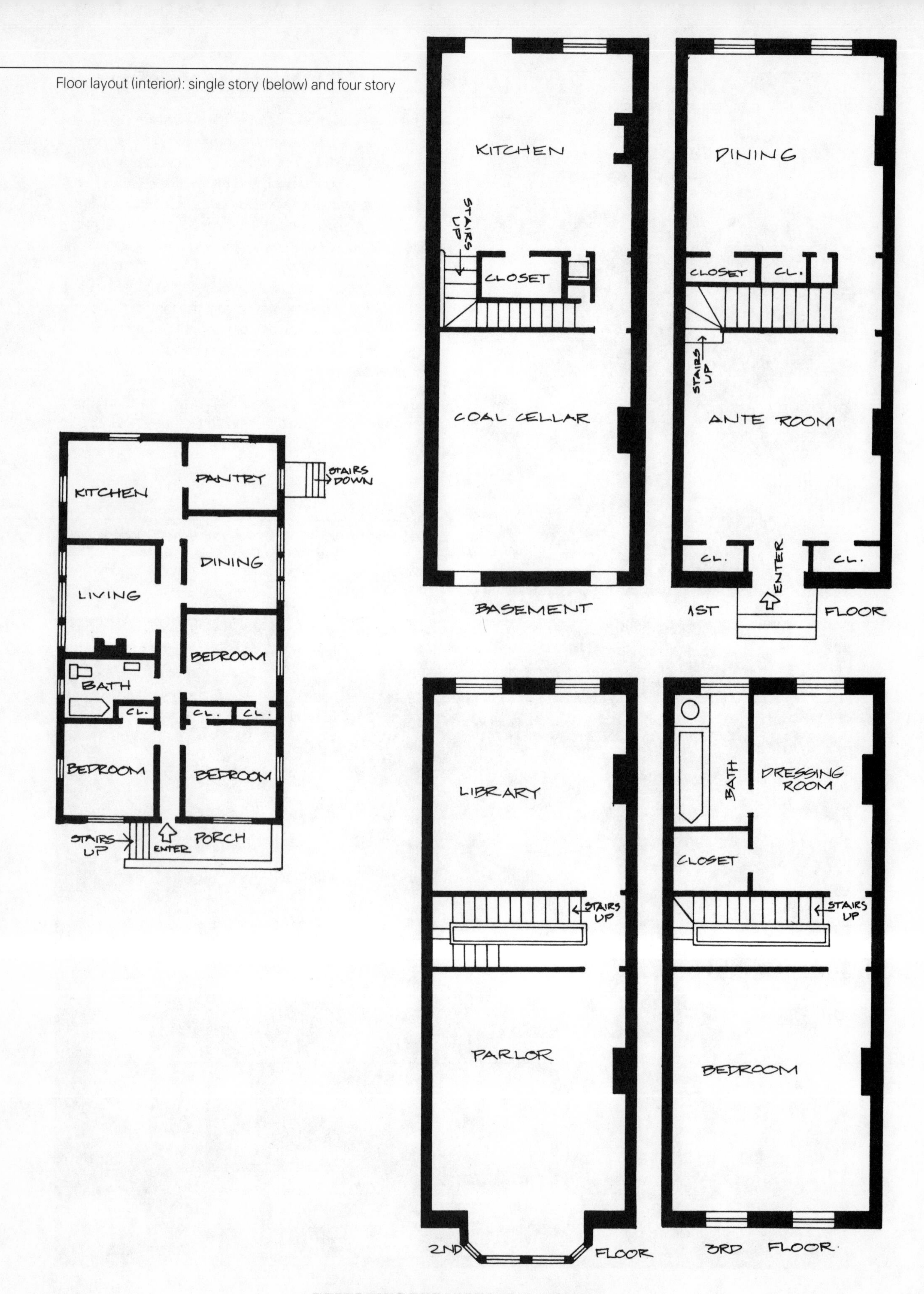

(RESEARCHING YOUR HOUSE)

It is fun to know who built your house and when, and what the original style looked like, both inside and out. The latter will also help you in restoring the house to its former self, as far as possible. Sometimes you can get information from former owners or neighbors in the area, but oral history is generally less than reliable. It is far better to get documentary evidence if it is available. (Records of homes before, say, 1860, would be vague or nonexistent in most cities.)

The building department of your local city is the best place to discover facts about your structure. If building permits were in force, you will be able to get these to tell you the name of the original owner, date of building, types of heating, and floor plans. For building department data you will need your block number and lot number. These are available from a plot book or from maps in city archives.

Your local historical society may also have knowledge of your Victorian — there may be photographs and other pertinent information to assist you in determining the authenticity and dates of your home.

This brownstone in New York has much of the Italianate style. (Courtesy of Landmark Preservation Commission)

Floor layout (interior): first floor of two story suburban

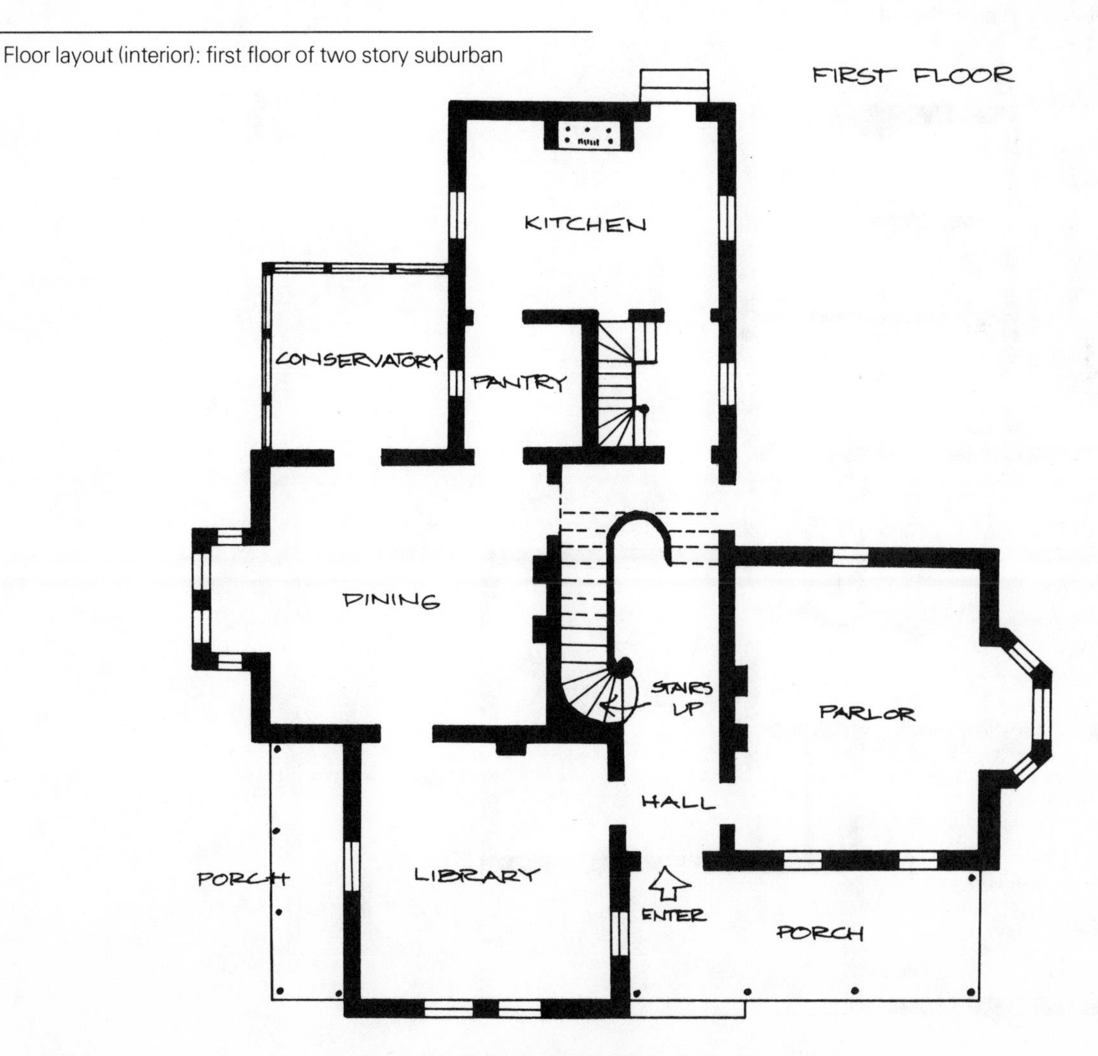

If all the above methods fail, you will have to search out information from the house itself. Old houses speak volumes—in their materials and more so in their design and style—and these elements can give you an approximation of the house's age. Look to windows, doors, chimneys, roof lines, porches, interior ornamentation and mantels to give you clues. They are there. Read them carefully; you generally can date a house within 10 to 15 years.

Physical evidence such as nails used, latches and hinges, sash and glass, can be used to help you guess the age of your home. Wire nails did not come into existence until after 1850. Moldings that are part of the door date back to the 1840s. After that time moldings were separate units attached to doors and known as strip moldings.

Knowing the age of your house is vital in making satisfactory restoration, so wherever possible do put all the clues together and be knowledgeable about your Victorian.

Floor layout (interior): second floor of typical suburban

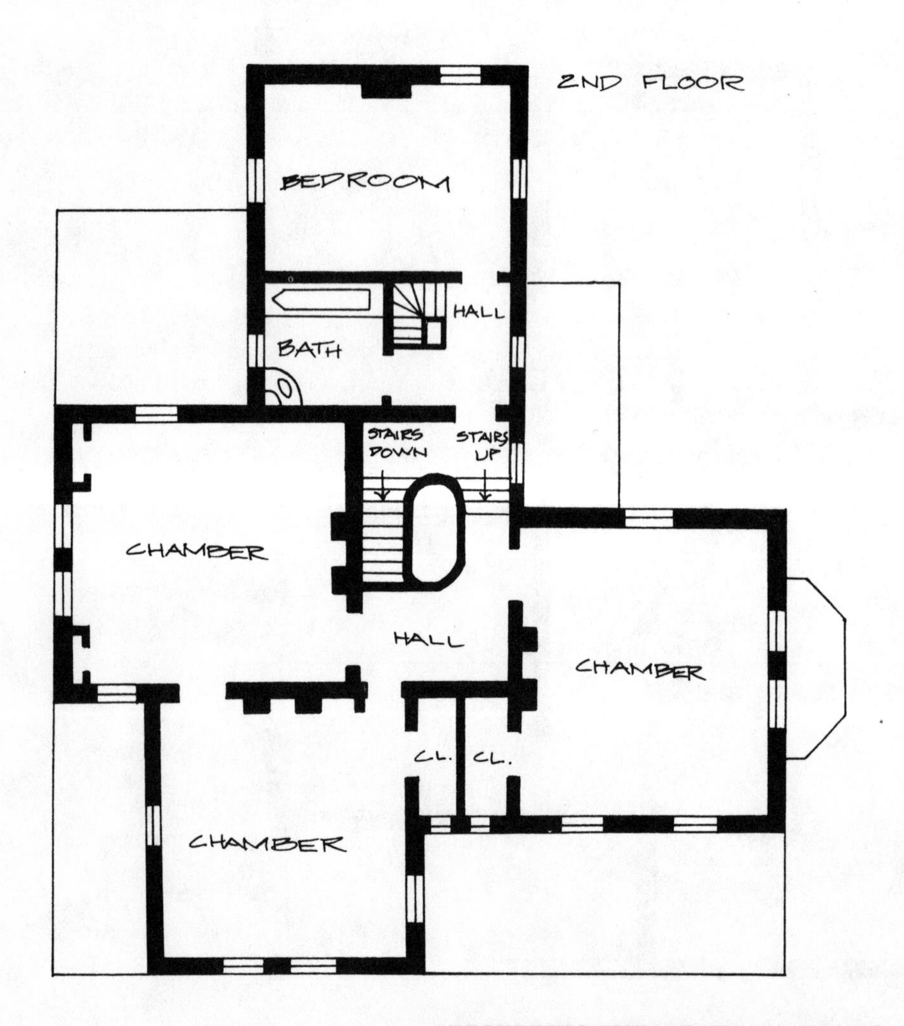

[CHAPTER 2]

EVALUATING THE STRUCTURE

Just what makes a building Victorian? Some people think the keys are the porticoes, others opt for the turrets or lookouts, and still others look for the rich adornment that frames the windows. But there is more to it than that. The structure itself—its balloon framing; the layout of the rooms, with double parlors; the long stairways; and the vertical character, all heightened by elaborate decoration—is Victoriana to most people. As works of the building trade, Victorian houses deserve to be historically preserved. The homes are a heritage of many large cities, where numerous Victorians were built; these homes became "the" house. This heritage is as strong a part of our history as are our contemporary skyscrapers and buildings, and as such its value to America is limitless.

No matter how you yourself define a Victorian house, if you have or want one and want to restore it, there are certain items you must check, repairs you must make, to turn your Victorian into a treasure as authentic in detail and construction as it was when first built. You must consider costs and local zoning requirements and codes. You will have to know what suppliers carry details like a witch's cap and the necessary materials, and you will also have to know when such items are not available commercially (in which case you can make your own or hire someone to construct what you need). Here we introduce you to the points you must consider when buying a Victorian and restoring one. Each structural point is fully elaborated upon in subsequent chapters.

[STRUCTURAL ANALYSIS]

The term *structural analysis* sounds highly technical, but it is merely an inspection of the foundation, roofing, electrical and heating systems, and plumbing of a building. This is done to determine if the structure is sound (and if so, how sound) and to assess how much work is necessary to restore it to safe operation. Once you have a knowledge of structural problems—and any Victorian is bound to have them—you can intelligently estimate just what it will cost to restore the building. Inside, you have to be sure load-bearing walls are sound and ceilings intact, or, if they are not, you should know just what is wrong with them. You have to look at sub-flooring and flooring to see how much work is needed in these areas. You have to inspect walls for cracks, hairlines, and so on.

Outside, check posts and piers, vents, crawlspaces, porches, railings, overhangs, columns, sheathing and siding, and, lastly, ornamentation. To a great degree you can always replace missing or lost ornamentation, but essential structural elements such as siding and columns are costly to restore and require expertise.

So, both inside and out, inspect slowly and carefully. For insurance, hire an architect (or someone well versed in building) to accompany you to see if yesterday can be salvaged and saved. Some Victorians may be well beyond the point of saving—only an architect can really tell. Generally, fees for this service are not too expensive and can be charged at an hourly rate.

[COST]

Do *not* spend more than you can afford. When you tender your bid to buy the house, include with the loan from the lending institution the construction money you will need. Shop around for rates because they often change weekly. Try to strike the best deal for your money, which means points and percentages paid for the loan itself.

For example, if you are about to buy a Victorian for $70,000 or $80,000, but it needs an additional $30,000 for basic rehabilitation, the lending appraiser should figure that the house is worth at least $100,000. If you get your home loan separate from your construction loan, your monthly payments will be larger than if you get both loans at once. Tell your broker you want to rehabilitate the house, and let the broker do some talking for you at lending institutions.

In some cases, cities themselves help rehabilitate old houses and offer special percentages and incentives. This policy differs in each region, so you must check out this matter with the proper organizations in your town.

(REBUILDING)

Can you tackle the restoration yourself? Yes, and you should try because it is challenging and exciting to restore a structure to its original concept. In fact, it is rewarding, so even if you are only halfway good at carpentry, give it a try. (You can always find knowledgeable friends who can bail you out if you run into major problems and by doing it yourself you will of course save money.)

However, you should hire professionals if the job requires skills you do not have or you want the job completed quickly. Sometimes local codes may force you to hire a licensed contractor, or the loan contract may stipulate that you do so. Professional help involves civil engineers to evaluate structural conditions, architects to design structural changes, and licensed contractors to perform the actual work. (To meet codes, plumbing and electrical work has to be executed by professionals.)

This may sound formidable, but do not let it intimidate you. You will be surprised at how much you can do yourself once you start. And you can probably do much more of the work than you think. Also, there are certain esthetics involved in dealing with Victorians, and you have a sense of this or you would not be involved. But the contractor you hire may not have this sense, so do try to do most of the work yourself.

(FOUNDATIONS)

The foundation is a prime element in the total Victorian picture because it takes the weight of the building and distributes it evenly over the soil. When working with a Victorian, you must know your foundation. Many of the homes built between 1840 and 1870 had no specific foundation work; they were built on mudsills placed directly on the ground. Framed walls were attached to the mudsills. Obviously such a foundation will need extensive replacement. Brick foundations were used in the 1860s, and concrete was introduced about 1910.

In old buildings, settlement of the earth is a serious problem. It causes shifting, and the foundation responds to the movement. Settlement, which can cause cracks and fissures in the foundation walls, can result from flooding, movement of the earth, insufficient drainage, rotting, and a small or large landslide.

INSPECTING FOUNDATIONS

As mentioned, if your structure is on a mudsill, complete foundation replacement is necessary. If the foundation is brick, look for these signs of potential trouble: loose and powdery mortar, crumbling bricks, cracks running diagonally across the brick wall. Allow sufficient funds for making all necessary repairs.

In a concrete foundation, look for cracks, which denote uneven settlement. Large cracks indicate that the foundation is moving — and thus the house. Smaller cracks or hairlines mean only minor settlement and could stay that way for years; it *may* take 20 years for a hairline crack to become a major hazard. However, even a hairline crack, if not repaired, can lead to new foundation work.

You should call in an engineer to evaluate the cracks and decide how much movement has occurred. You could consult a contractor, but remember that contractors make their money from repairing foundations, so a contractor's advice may not be as objective as an engineer's.

You can inexpensively fix chipped concrete corners yourself. Minor cracks in concrete can be easily repaired; major ones require more repair and consideration. A new concrete foundation for a 1500-square-foot house can cost as much as $2,500. In addition, if you have to shore up the building to permit the work, you have to add another $5,000.

Most pre-World War II homes are supported by wood posts on concrete piers underneath the building. Inspect posts and piers for dry rot, termites, and deterioration of any kind. Post and pier repair work is not as extensive as foundation work, but some cost is involved, whether you do the repairs yourself or have them done.

Sill plate damage and anchor bolts are other things to look at and for when inspecting the foundation. The *sill plate* is a horizontal wood piece between the house frame and the foundation itself. If there is any damage to it, such as decay or termite infestation, it will have to be replaced, which is not easy to do. *Anchor* and *seismic bolts* fasten the sill plate to the foundation and help stabilize the building. Seismic bolts are especially important in earthquake country or where there is any sizable settlement of the earth. They are secured into the mortar of brick foundations, directly into concrete ones.

(CRAWL SPACE)

The *crawl space* is the space in buildings without a basement between the ground and the bottom of the first floor. It is an important component of a building because it provides air to the bottom of the house. Check for moisture retention, beetles, termites, and be sure there is sufficient clearance between the ground and the wood members of the house. For maximum safety of the structure, there should be a 6-inch clearance between the ground and the wood sill that sits on the masonry foundation. Look for at least 18 inches of clearance between the ground and wood members (floor joists) and 12 inches under a beam supporting the floor joists. Be sure there are 6 inches between wood posts that sit on concrete piers and the ground.

Check for vents, which are rectangular openings around the building that provide air circulation and help retard moisture retention. They should be situated every 150 square feet of ground area and ideally opposite each other to provide good cross ventilation under the structure. Most vents in Victorian homes are made of ornamental cast iron.

[ROOFING]

A house is as good as its roof. A roof that does not leak is essential to the life of the house. Once damage starts from roof leakage, it can deteriorate exterior siding and interior finishes, so you must inspect the roof carefully. Siding and ornamentation can be replaced with little work and at relatively little cost, but structural components like walls are much more expensive to replace.

Before buying your Victorian, go up on the roof and inspect the roofing material. Look for protruding nails, sagging ridges, and missing or worn-out shingles or tiles. Also thoroughly check the flashing at ridges and chimneys. *Flashing* is a strip of thin gauge sheet metal that prevents water from entering the roof. It covers exposed joints and diverts the water to other areas. On old buildings, flashing is copper or lead, and invariably it is in bad shape. Look for flashing at the ridges and hips of the roof, in gutters and eaves, and in door and window openings.

If you can get to the attic, watch for water stains—always a clue to trouble—and warped or sagging rafters. Check ventilation. A properly ventilated attic space will prevent rot and structural damage to the rafters and will let moisture evaporate. Vents, important to the roofing skeleton, should cover 3 percent of the ceiling area of the room below.

No doubt any old building will need some roof repairs, but the idea is to determine before you buy just how much repair will be needed. Often a complete new roof is in order; other times minor leaks can be repaired with asphalt roofing compound and liquid roof coating. For a new roof you want a durable one within your budget. You can select from various materials, including asbestos siding, wood shakes, shingles, and roll roofing. (We look at these materials in Chapter 5.)

[SHEATHING]

Sheathing (siding) is a weatherproof surface applied over the exterior of the wood structure to protect it from the elements. It is often in disrepair on older buildings, and matching existing siding can create problems because it may no longer be made. There are several kinds of wood siding: horizontal, consisting of drop siding; shingles; bevel siding; clapboard; and vertical siding, which is generally board and batten and is often found on Stick Victorians. Chapter 5 describes sheathing in greater detail, including wood shingles.

When you inspect siding, look for cracks and decayed wood. Use the pick test, as for porches, to discover any rot. Test with a screwdriver or a sharp key. Small fissures are not a serious problem and can be remedied with putty or caulking. Wide cracks are more serious but can also be repaired with appropriate caulking materials. Warped boards are quite serious and take more fixing.

Wood shingles are flexible, 16, 18, or 24 inches long, and made of redwood or cedar. They are octagonal or diamond shaped and can be formed into various patterns to decorate the facade.

[WINDOWS]

Windows are the eyes of a house and are vital in the rehabilitation of the Victorian. They should be replaced with necessary authentic duplicates. You may skimp somewhere else, such as with ornamentation or posts, but with windows you must stay as close to the original as possible if you want to preserve the character of the structure. (See also Chapter 6.)

Generally, windows of Victorians were double-hung, casement, and fixed. A double-hung window opens with an up and down motion and has two panels: upper and lower. It was and still is a sound window. The window movement is controlled by chains or pulleys with weights or by a spring mechanism concealed in the side jamb. In time double-hung windows become askew as pulleys get fouled up or cords fray. However, these problems are easy to fix.

The casement window with its small panes is charming and graceful. A window opens outward; the window is attached to the frame by hinges on its vertical edge. These windows last for decades and need little repair beyond tightening the screws or replacing the sliding rods.

The fixed window is a window that does not open; it usually has leaded or stained glass. Leaded and stained glass windows may need repair after awhile because they become bowed, but this is not a serious problem.

(FRONT DOOR)

The front door of a Victorian is more important than you may think because it indicates what to expect inside the house. Heavy and solid oak doors with beveled or flashed glass on top and recessed molding below were popular on many Victorians. Unfortunately, the beautiful front doors of many Victorians have been damaged or replaced. If the door is damaged, do all you can to salvage it. If it has been replaced with a modern door, search salvage yards to find a true Victorian front door. This special part of the house must be authentic. (A more complete discussion is in Chapter 7.)

(STAIRS)

One of the most common trouble spots in Victorian homes is the stairs. Made of wood, they invariably show rot, and if there was any haphazard work done on the original building, it was probably in the stair area. Drainage is the main problem—any water that accumulates starts the rotting process. Look for lack of a protective seal like paint or caulking, and check for worn areas that hold water or solid boards installed perfectly flat, both causes of water retention and eventual deterioration.

Most Victorians have wood stairs, but occasionally concrete or brick has been used. Neither material is really true to the Victorian style and so should be replaced with wood. Also consider handrails, which are subject to deterioration and will probably need replacement. (Chapters 8 and 13 cover stairs and the interior staircase.)

(FRONT PORCHES)

The front porch, whether very wide or small, was an integral part of the Victorian, sheltering people from rain and providing an attractive embellishment to the house. The porch on your home must be restored authentically in terms of ornamentation and structure.

When inspecting the porch, look for any wood that (1) feels spongy, (2) shows splits or is flaking paint, or (3) is charred; these three factors denote rot. To investigate the extent of damage to the wood, poke the wood with a sharp tool: if it is soft, there is trouble. Also, remove a sliver of wood; if it lifts out easily, there is some rot; but if it splinters, the wood is still in good shape.

Check step treads for worn-out spots, and look at the joints in the railings. Inspect the underside of the deck and framing to see if it is sound, and always look at the bases of posts to be sure they are in good condition.

Many typical Victorian porches had flat roofs that collected water, leading to problems. Inspect such a roof carefully. (There is more information about porches in Chapter 8.)

(EXTERIOR ORNAMENTATION)

Ornamentation is not only an essential part of the character of the Victorian house, it is also the beauty of the style. Exterior ornamentation may be made of wood, plaster, or pig iron. Often, to repair any damage underneath, you must remove the ornamentation, and frequently the ornamentation itself needs repair. Before you rip off pieces, study them to see how they are attached so you can remove them with as little damage as possible. Use a prying tool like a small crowbar, and slowly and carefully apply pressure below nails to avoid cracking the pieces. (More instructions are in Chapter 9.)

(FLOORS)

In most Victorian homes, the floors can be repaired. Indeed, unless the house has severe structural problems, the floor is the easiest part to restore. You do not have to remove ornamental pieces or replace siding; all that is usually necessary are some minor repairs and sanding, because hardwood floors last for decades.

The subfloor is the base for the finished floor; in older Victorians the subfloor is usually wood planks nailed perpendicular to the joists. In more recent houses, the subflooring is installed diagonally, at a 45-degree angle to the joists. You can use plywood sheets as your subflooring. Most Victorian subfloors are in fairly good condition; although if joists sag, the boards may be loose and the finished floor might squeak if you walk on it. Removing a subfloor and putting in plywood sheets is not a difficult job.

The finished floor is the one you see and walk on. In most cases it matches the ornamentation and character of the house. Starting about 1890, wood became the choice floor material, and today we still have beautiful plank or parquet floors in many Victorian houses. Hardwood was reserved for living and dining room floors because it is incredibly durable; softwood floors were used in bedrooms.

The typical Victorian wood floor is oak, 2 inches wide by $\frac{5}{16}$ inches thick, but there are many variations among the parquet floorings. Loose boards can be easily repaired, and protruding nails that loosen over time can be refastened. The occasional cracks in Victorian floors are minor problems.

There should be little problem with the structure of the flooring itself, but refinishing is almost always necessary. You have to sand the wood down to its natural state and then protect it with a coating like polyurethane or a color stain. (Refinishing floors is discussed in Chapter 10.)

[WALLS AND INTERIOR WOODWORK]

Through the years walls invariably develop cracks, holes, or other odd defects. At first glance the walls may seem beyond repair, but they are not; wall repair is not that difficult to do, as long as you are patient. Most Victorians used plaster as the wall surface. Plaster is a dough-like material that hardens after it is applied over a flexible framework; as a house settles, cracks do develop in the plaster. However, you can repair the cracks and chips with modern adhesives. Very large cracks or holes have to be replastered (which requires someone who is quite knowledgeable about wall repair) or covered with sheetrock. Sheetrock or gypsum board, 4 x 8 feet, of varying thickness, is like a solid sheet of plaster; a plasterlike paint is applied over it to create a wall that looks almost like authentic plaster.

You can fill hairline cracks with a mixture of plaster or joint cement. You must open and undercut small cracks to create an opening. Then fill the cracks deep (about 2 inches from each side) with a mixture of plaster and joint cement.

The decorative wood in Victorian houses is the main reason why the old houses have warmth and ambience. *Wainscoting* — a skirt of wood on the lower three or four feet of a wall — is quite common and should be restored. Other wooden ornamentation, from ceiling beams to columns and moldings, should also be preserved.

With time and care you can strip and refinish any wood. Replace damaged wood pieces. (Lumber yards have a wide stock of interior lumber.) If you cannot match the wood exactly because it is no longer available (for example, gum wood), use a suitable substitute to look like the original. (Chapter 11 is a full discussion of wall surfaces and woodwork.)

[CEILINGS]

The height of the ceiling (Chapter 12 elaborates this discussion) is an important feature of a room because it creates a certain feeling. High ceilings, most often found in Victorians, denote formality and elegance. *Never* change the ceiling height, because it is an integral part of the Victorian charm; ceiling damage can be repaired. A plaster ceiling is prone to cracks because of settlement of the building and water leakage. The same plastering techniques for walls can be followed for ceilings. If the ceiling is sagging, press the surface back into place by drilling a hole through the plaster and installing a ceiling anchor.

If there is extensive ceiling damage, replaster the ceiling or use sheetrock. Once the sheetrock is taped, you can use one of the textured paints that impart a look similar to that of old plaster.

A *rosette* is a decorative support for a chandelier. If there is a rosette in the center of the ceiling, try to restore it. Rosettes vary in shape and size, but all are highly decorative. New rosettes are available from suppliers if the original rosette is too far gone.

Retain ceiling moldings and other trim, including coved designs, which give a room a special feeling.

[DOORS AND HARDWARE]

The interior doors of Victorians are invariably handsome and superior to those manufactured today. Most doors have some ornamentation that is appropriate to the overall design of the house, so try to retain and restore old wood doors rather than replace them. If the door is binding or loose, it can be repaired inexpensively. (More about doors in Chapter 13.)

Likewise, retain all hardware. Beautiful brass and porcelain hardware knobs and hinges were used extensively in older homes. If some are missing, replace them through the many suppliers who now specialize in Victoriana (see list at end of book).

[UTILITY SYSTEMS]

When you come to the plumbing, electrical, and heating systems of your Victorian it is almost a sure bet that most of these systems will have to be revamped for modern conveniences and today's living. In fact, a major portion of the renovation of a structure is usually concerned with these major facilities. While all systems may still be working, they are generally outmoded and must be brought up to code in most areas and this could involve extensive repair and money. But generally in any old house you buy, this is the case. So do not avoid a handsome Victorian on the merits of its utility systems — they can be repaired and structured to today's standards and much of the work you can do. We explain how in the last section of this book.

[CODES AND PERMITS]

Each region has its own building safety codes that you must follow. Getting permits for remodeling work is necessary as well. Check with officials first before you start; if you begin a restoration without a permit, you can be made to stop and dismantle it or be fined heavily.

Housing codes involve many factors, such as fire-resistant walls between the building and the attached garage, two ways of egress from the top story to the second story of houses with three or more stories, and so on. Be aware that before any project can go ahead

legally, the permits must be acquired. The permit states that work will be done in compliance with existing building laws. Generally, you need a building permit, an electrical permit, and a plumbing permit.
And do not forget termite inspection reports (where applicable) before you even consider an old house. Some of these can knock you off your feet. Recently we bid on a 1901 Victorian at a price of $55,000. But the building had $35,000 worth of termite damage!

This book instructs the homeowner as to how to go about the structural and aesthetic restoration of the house. You must remember the *legal* aspects: all construction, electrical, plumbing, and mechanical work (other than minor repairs) must be approved by the city or county *prior to any work being done.* This entails obtaining the respective permits. Permits ensure that a clear and defined plan has been designed and that all intended building and installation adhere to the uniform codes of the particular crafts as well as the local ordinances of the city or county.

People often protest and shun code compliance. This is really not necessary. The standards established are not an act of fascism, nor are they meant to be punitive in any way. Codes are merely a set of minimum requirements that must be met to best ensure the house's stability, safety, comfort, and function and to maximize its safeguards against fire, collapse, flooding, hazard, and contamination. If there were no guidelines and inspection and approval agencies, unknowledgeable individuals could attempt something seemingly workable and safe but in reality full of so many technical shortcomings that living in the house would be a risk.

Familiarize yourself with the uniform codes of the various trades, and take advantage of your local Building and Public Works departments. They are there to assist you in your efforts to restore your house, and they will answer most of your questions.

(CHAPTER 3)

FOUNDATIONS

A sound foundation is the basis for all restoration work because a weakened foundation threatens the very structure of the house and all its valuable architectural features. As mentioned in Chapter 2, the foundation sustains the weight of a structure, distributing it evenly to the ground below, and provides firm anchorage between the building and the soil.

(TYPES)

There are several different foundation designs. In the United States and similar latitudes, there are four basic types; their use is dictated by climate and soil conditions.

FROST-LINE CONSTRUCTION

Throughout parts of the eastern and north-central sections of the country, winter temperatures are cold enough to freeze the ground to a depth of 6 feet or more. To maintain a solid, stable connection to the ground, the house "anchor" must extend below the frost line. Hence, the front porch and stairway leading up to the entry level have become architectural characteristics of this region.

SEISMIC AND HURRICANE CONSTRUCTION

Areas afflicted by high winds or earth tremors must take precautions against the literal disconnection of the house from the foundation, and, in the West, must lessen the potential of the foundation crumbling altogether.

Throughout the South and Southeast, where both flooding and hurricane troubles are present, combination procedures must be taken. The best remedies for the winds are steel seismic and hurricane foundation anchors (available at hardware stores and building suppliers). This hardware prevents the building's framework from hopping off of its base.

In the West, neither high winds nor flooding are prevalent dangers, but earthquakes are. Foundations must thus be steel reinforced to a greater extent than in any other portion of the country. The western climate accommodates almost any type of foundation, but the most common are the slab, wall and footing, or combination wall and footing (post and pier). Reformed steel reinforcing rods must be placed in every poured concrete or concrete block wall. In slab-type foundations, crisscross squares called *wire mesh* and *footing rebar* must be embedded into the material.

By nature, masonry has an incredible amount of compressive strength and can, without interruption, withstand quite adequately the weight of a building. However, the downward force of a building is never the sole stress upon a foundation; shifting, irregular earth settlement (as discussed later), and seismic forces create horizontal tensions and lateral shearing impacts. In seismic zones, the tensile strength of steel reinforcing and foundation anchors is the best precaution against foundation walls pulling apart and buildings dislodging from them.

UNSTABLE SOIL CONSTRUCTION

Where soil is unstable for a considerable depth, such as in fill, marsh, or sand areas, it is economically impractical to excavate to a depth of several stories to anchor a foundation well to solid ground. *Pilings*—long cylindrical steel and concrete shafts that act as stilts—are driven down through the unstable soil and firmly affixed to the solid ground below (often bedrock). The pilings provide enough lift and stability so that a floor framing system can be placed on top of them.

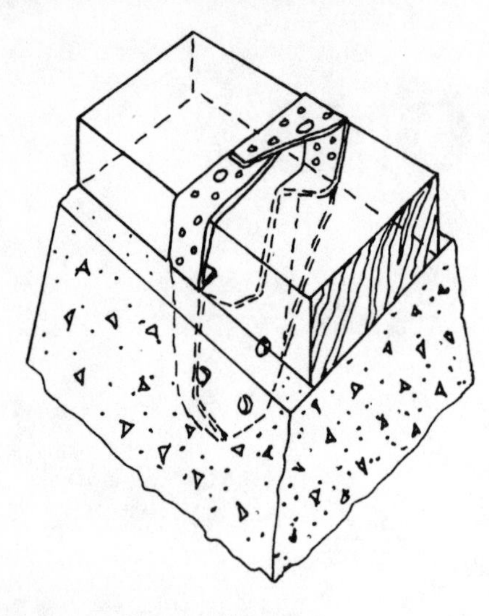

No-float foundation anchor bolt

Another remedy for unstable soil is the *raft foundation.* This is a floating bathtub of concrete and steel, with enough bouyancy and lateral stability to sustain the weight of a structure.

Neither piling nor raft foundations are in widespread residential use because of the expense and substantial engineering required. Such construction is limited to commercial and industrial structures.

WEST COAST VICTORIAN FOUNDATION

The West Coast Victorian was almost always constructed with a combination wall and footing — post and pier — foundation. This type of foundation has a short (about 6 inches to 2 feet high) exposed perimeter wall of concrete, stone, or brick. The dirt area enclosed by the continuous wall is uniformly spotted by a row or rows of square wooden posts resting on concrete blocks (about 12 × 12 × 16 inches high). On top of these posts, spanning from one end of the foundation wall to the other, are large, heavy beams called girders, which carry the weight of the portion of the house between the exterior walls. In wood construction, spanning great distances without support underneath would cause the floor joists to sag and break at the center.

In some houses, the supporting posts are high enough to create a space below one-half to three-quarters of the height of the rooms above. A compensating vertical stud wall is then built on top of the foundation wall. A continuous sill of boards resting on top of the foundation is anchored firmly to the masonry, on top of which is the stud wall. The floor framing for the living space above is constructed between the exterior stud walls and interior girder supports.

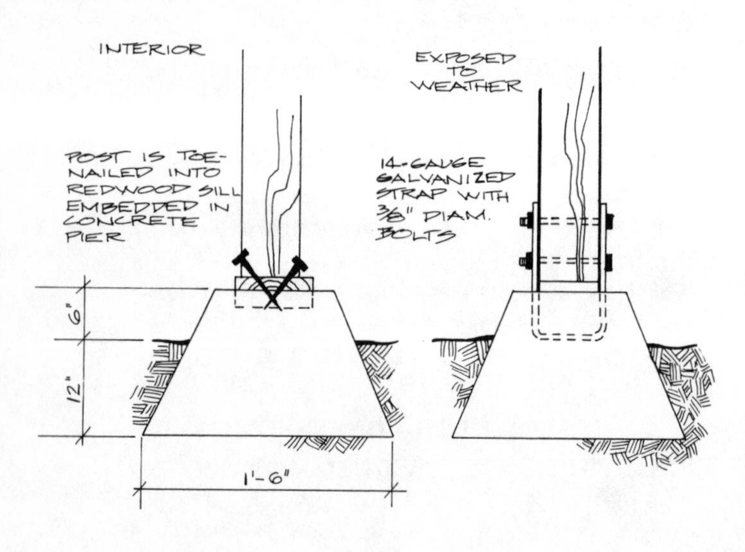

Post and pier elevation

[SETTLEMENT PROBLEMS]

Settlement is the major threat to a foundation. When the ground beneath a building moves, the foundation responds to the movement by adjusting to the new lay of the land. Because solid masonry is not flexible, the change in position causes cracks in the foundation walls. As these imperfections become larger because of more movement or natural erosive forces, the foundation loses the strength it needs to support the rest of the house.

No structure is immune from the effects of settlement. For newer homes, construction procedures take into account the initial settlement and orient and adjust the buildings to the environment. Older buildings, however, have been subjected to years of natural wear and tear as well as manmade impact.

Foundation cracks can be caused by the following seven soil settlement problems:

INADEQUATELY COMPACTED FILL

Variations in a soil's stability will cause a house to sink a bit into the looser soil.

NATURAL, GRADUAL EROSION

Because of the influence of rain, wind and sun, the soil beneath and around a building will take the course of least resistance. Rain and wind will often drain and sift out certain soil components from the ground, depositing them elsewhere, and the sun will dry out the soil, parching and powdering it.

NATURAL, ACUTE EROSION

Many homes are fortunately spared from flooding, landslides, and earthquakes, but some are not. Flooding and landslides can be forceful enough to literally knock the foundation out of place or completely detach it from the house. And seismic eruptions can cause the earth to settle irregularly in a matter of hours or days, therefore subjecting the foundation to extraordinary settlement shifts.

EROSION FROM NEARBY DEVELOPMENT

Improperly diverted drainage is a problem where homes are on sloping land. All too often water erodes the soil from around a house's foundation and deposits it in the backyard of a neighboring home. The home on the higher elevation suffers from the gradual soil decay, and the lower-level home is faced with a continuous build-up of pressure along one of its sides.

POOR CONSTRUCTION

Many people, attempting a quick patch job on their foundation, overlook the essential ingredients of time and quality. If there is dampness and water seepage in the basement area, they simply stuff any holes and throw dirt on the water. They patch cracks up with any mixture that simulates mortar in texture and color. They wedge bricks back into a level position and then plant shrubbery around the house to hide these flaws. These easy solutions work fine for real estate agents and owners who want to sell a visually correct package. However, all these "remedies" actually worsen the damage caused by settlement, permitting it to amplify while going untreated.

ROTTEN TIMBERS

Wood is an organic material that gradually decomposes; such natural deterioration can take years and years. However, through moisture, drying, and termite infestation, wooden support systems can quickly weaken and rot. Sagging floors and bowed walls are usually the first signs of this problem. When these symptoms are left unattended, the weight of the house will shift and depend more on the perimeter foundation walls. These walls, not designed for excessive lateral force, will shift and crack where the pressure is the greatest and the soil the least stable.

ROOTS

The soil around and within the confines of the foundation wall should be poisoned, to prevent vegetation from cropping up in critical areas. Where trees grow close to the house, the roots may grow under the foundation wall. As unbelievable as it seems, tree roots are incredibly strong and persistent and will push a concrete wall out of place if given a chance. The results range from running hairline cracks to the advanced stages of a wall's being cracked in two, to permit room for the roots to pass through.

[DETECTING AND REPAIRING DAMAGE]

BRICK FOUNDATION

A brick foundation is a rare and valuable element of antiquity. You should make the best possible effort to respectfully restore and maintain its beauty and character. When inspecting a brick foundation, look for the following five trouble signs:

1. Cracks between the bricks, especially cracks running diagonally across the wall. This indicates that one or more settlement problems is present.

2. Loose and powdery mortar. (Scrape the mortar with a penknife or screwdriver; the mortar should be firm and solid.) Cakey mortar is usually the result of age and exposure to wind, rain, and sun.

3. Crumbling bricks. This problem is usually caused by exposure. Left untreated, such bricks will severely impair structural stability.

4. Bricks that do not run in a true horizontal line from corner to corner. This is another indication of some kind of settlement problem.

5. If the brickwork is severely cracked, crumbled, or out of line with the horizontal, consult a civil engineer to determine the exact nature of the problem and how to best remedy it.

Many local codes no longer allow brick foundations as structural systems. Many experts feel that a brick foundation's lack of steel reinforcing disqualifies it from providing lateral stability. Thus, if 75 percent or more of a foundation needs major repair, many cities and counties will require that a new code-adhering foundation replace the old one. The recommended new construction is usually concrete—block or poured—with steel reinforcing. If this is the case in your situation, do not cry or panic; salvage the old bricks and reuse them as a veneer.

Repointing. However, if the brick foundation is just slightly worn and shabby, you can repair it by repointing, which is a process of patching up cracks and crumbling mortar. To repoint:

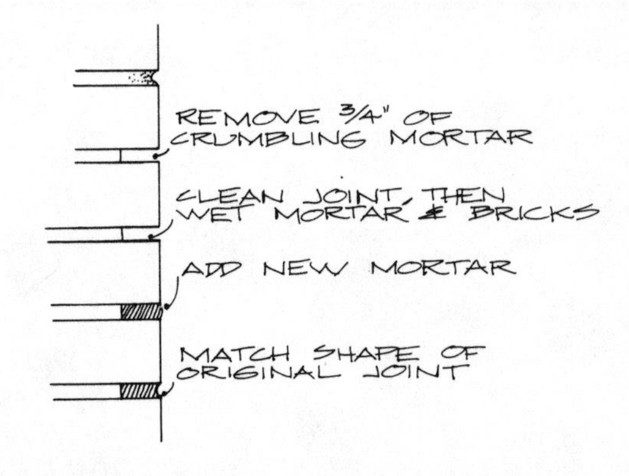
REMOVE 3/4" OF
CRUMBLING MORTAR
CLEAN JOINT, THEN
WET MORTAR & BRICKS
ADD NEW MORTAR
MATCH SHAPE OF
ORIGINAL JOINT

How to repoint

1. Remove the old surface mortar to a depth of ¾ inch. Be sure to clean out all the loose mortar and clean off all dust.

2. Dampen the joint to keep it from drying out prematurely; apply mortar so it fills the space of the original mortar.

3. Smooth the mortar with a trowel, making the new mortar joint match the old one in size and design. This may require a special striking tool.

Select a mortar that is the same color and strength as the original. The new mortar should be slightly darker than the old mortar when both are wet because the new mortar will lighten as it dries. Vinyl cement, although easier to apply, never matches the original mortar. Consult building and masonry suppliers; these experts can give you the best recommendations for mortar mixes and equipment.

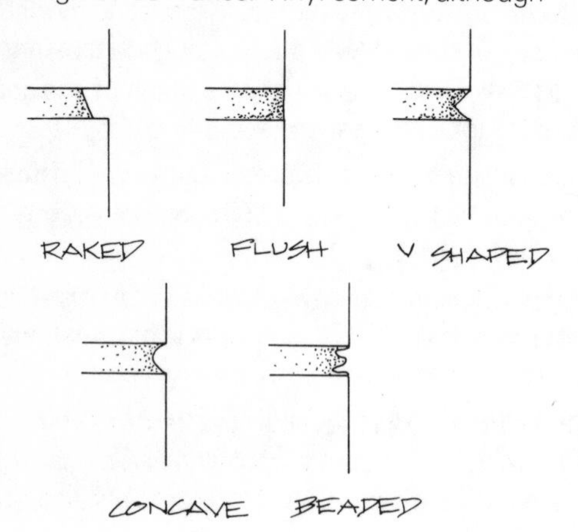

Typical mortar joints

Replacing Bricks. When replacing bricks, be sure to use bricks that are the same length, width, height, and color as the original ones, and by all means install them in the same pattern. If this is not done, all your toil and effort will appear to be nothing more than an obvious patch job.

Clinker Brick. Classic houses are known for the clinker bricks used in the foundations, fireplaces, and chimneys. These are irregular, unevenly fired bricks that manufacturers used to discard as rejects. Although clinker bricks were available as late as the 1950s, the only way to find legitimate ones today is through a wrecking company. If authentic clinker replacement is impossible, consider these two reasonable alternatives:

1. Flashed brick. This brick is geometrically shaped, like a conventional brick, instead of being gnarled and pitted like a clinker. However, it is chemically treated, to show dark stains like its original counterpart.

2. Broken brick. You can find this brick in the reject pile of most brickyards. It is a typical red brick that broke during some stage of the manufacturing process. Although the color does not match that of the clinker, it is pitted and irregular in size and shape like its ancestor.

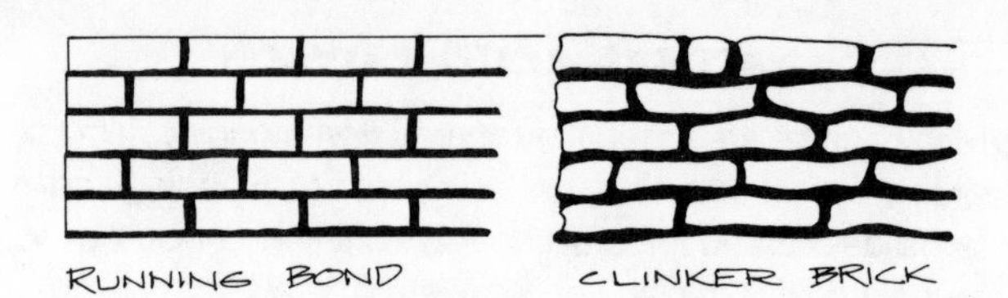

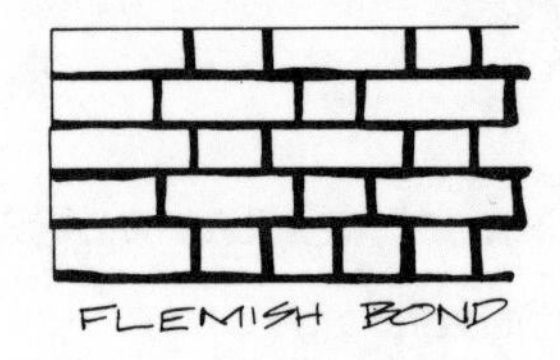

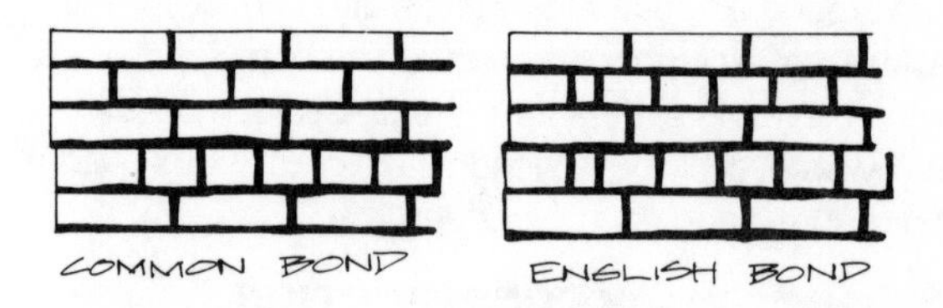

Brick patterns

CONCRETE FOUNDATION

Most homes built after the turn of the century have concrete foundations. These foundations are subject to erosion or settlement problems. The obvious evidence of uneven settlement is cracks in the foundation wall, and

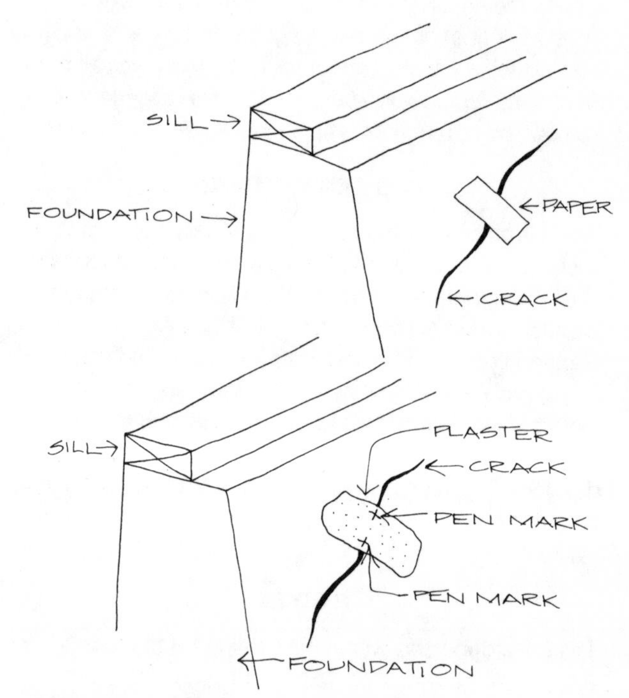

SILL
FOUNDATION
PAPER
CRACK
SILL
PLASTER
CRACK
PEN MARK
PEN MARK
FOUNDATION
PLACE A BRITTLE MATERIAL, SUCH AS GLUED-DOWN PAPER OR PLASTER OF PARIS OVER CRACK. IF PAPER TEARS, OR IF PLASTER CRACKS WITHIN 2-4 WEEKS, SERIOUS PROBLEMS MAY BE INDICATED.

Testing for movement in foundation

Replacing a portion of the foundation

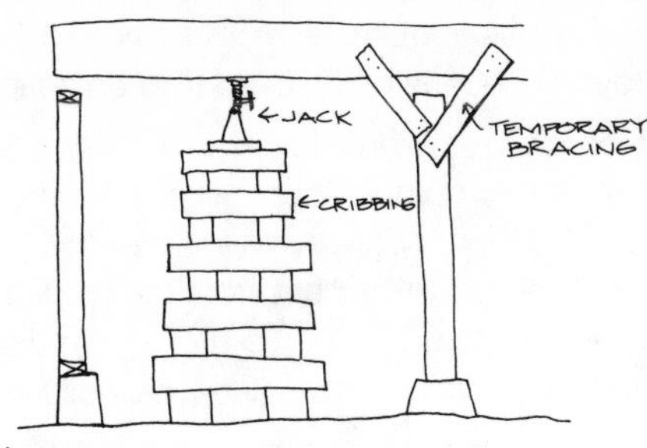

1) PLACE PORTION OF HOUSE ON JACKS.

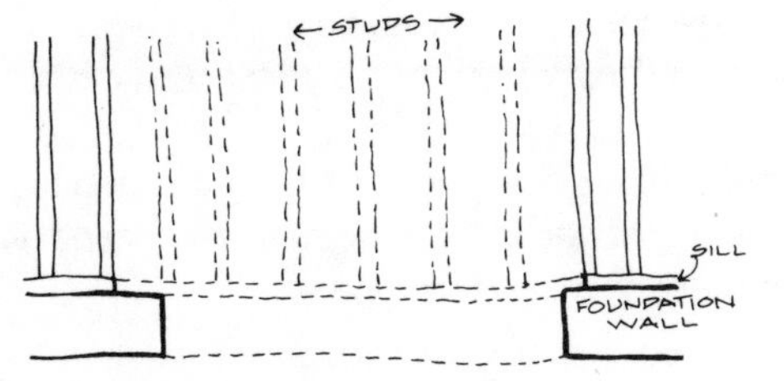

2) REMOVE DAMAGED PORTION, PLUS ADDITIONAL LENGTH.

CUT BACK FOUNDATION SILL TO PERMIT LAPPING OF NEW SILL OVER OLD FOUNDATION.

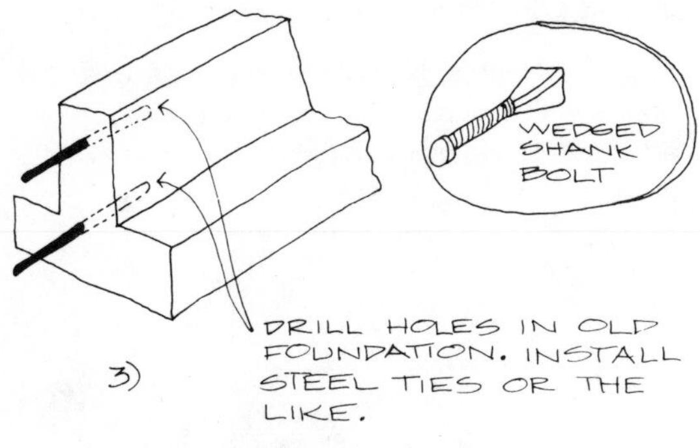

3) DRILL HOLES IN OLD FOUNDATION. INSTALL STEEL TIES OR THE LIKE.

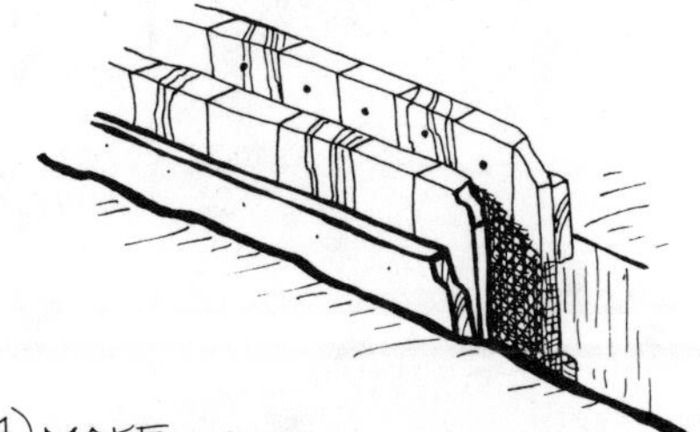

4) MAKE CLEAN EXCAVATION AND CONSTRUCT WOODEN FORM AS A MOLD FOR THE CONCRETE. POUR THE CONCRETE; LET DRY, REMOVE MOLD.

5) REPLACE SILL, AS SHOWN IN FIGURE "2". RECONSTRUCT STUD WALL.

6) LET HOUSE DOWN FROM JACKS. RE-ATTACH POSTS AND GIRDERS.

around door tops, sills, and window frames. To differentiate between minor cracks caused by irregular settlement and major cracks that are worsening with time, conduct this simple test on the foundation: with a contact adhesive, glue a strip of paper across the crack (or use plaster of paris with two pen marks instead of the paper strip). You want a brittle material, one not inclined to stretch. If the paper tears (of if the plaster cracks) considerably along the crack in the foundation in 2 to 4 weeks, you may have a degenerative settlement problem. Consult a professional engineer.

If the foundation cracks are minor, you can repair them by scoring the concrete with a sharp and strong tool, cleaning out any moss, dirt or spiders, and filling the cracks with concrete patch cement (available in small packages).

If the cracks are major, the repair job can be quite a project. The cracks may have separated the wall into two or more distinct elevations; this usually occurs after the flaw has been left unrepaired for some years. Nature takes its course by letting one side settle or one lift, or a combination of both. Simple patching cannot rectify the damage. The structural integrity must be maintained, and to accomplish this, expert knowledge must be applied. If you cannot do it yourself, make sure the engineer recommends something similar to the following eight suggestions. (It must be emphasized that these eight steps are only guidelines of what must be done. Every foundation, soil condition, and local building law

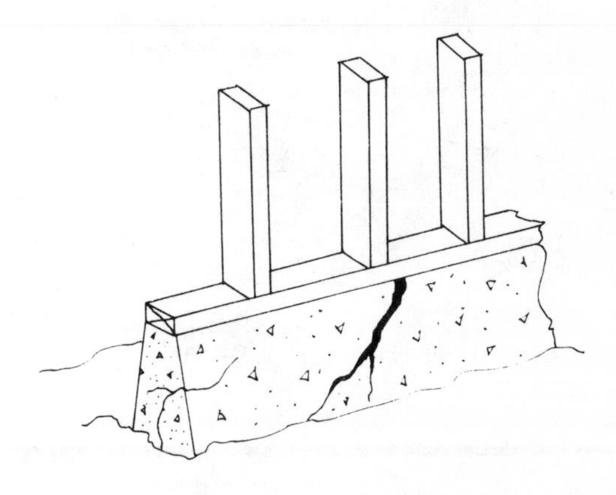

CRACKED FOUNDATION

differs enough so that each problem must be examined separately and the specific solution prescribed by a building professional.)

1. Place the damaged portion of the house on jacks. (See the instructions for jacking up a house in the following Post and Pier Foundation section.)

2. Remove the damaged portion to 3 feet beyond each nearest bearing line.

Repairing cracked foundations

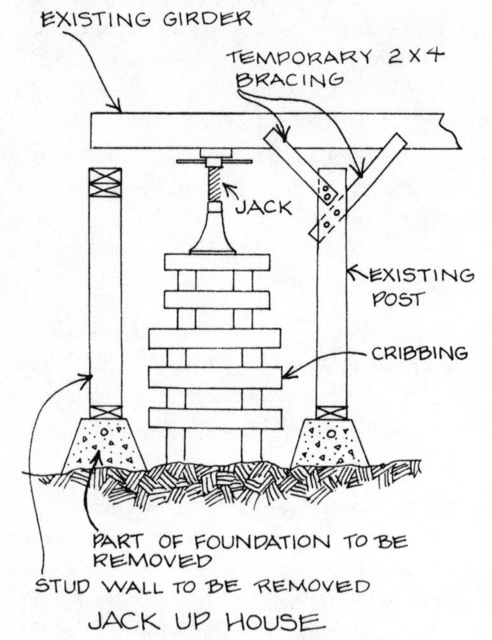

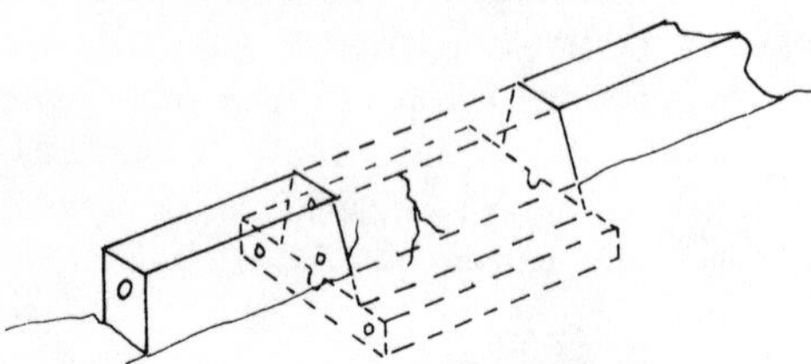

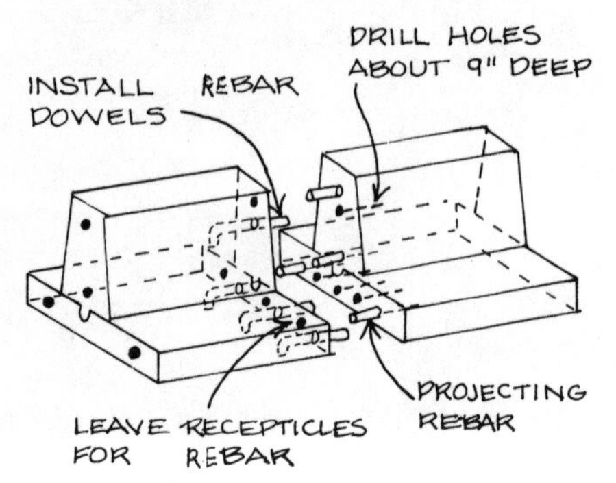

3. Drill $\frac{9}{16}$- or $\frac{13}{16}$-inch holes in the cutaway section of the old foundation (one hole 2 inches from the top of the wall and one hole 2 to 3 inches from the bottom of the wall). Drill the holes 6 to 9 inches deep, depending upon the type of foundation.

4. Firmly anchor ½- or ¾-inch steel ties into the old foundation. (This is difficult but can be done by several methods: hooking, bolting, or adhering them in some detachable manner.) Building suppliers sell a special wedge-shank bolt for this purpose.

5. Dig an excavation that is clean and uniform with that of the old foundation, but make it wide enough to fit the wooden concrete mold. Build the concrete mold form.

6. Place the required rebar as specified by the building code (usually number 3 or 4 type), 12 inches apart, two rows in the footing; two rows in the wall, 1 to 2 inches from the top; 1 to 2 inches from the footing line and vertical ties, 1 foot to 2 feet on center; depending upon the foundation characteristics and local building ordinances.

7. Pour the concrete into the mold with its network of rebar. Use ready-mix concrete, or mix it on the spot. For extensive jobs, a premixed type is best because it lessens the potential of mixing two or more separate batches that are of different consistencies. Also, you have to clean the barrel or mixer between batches, and during that time the wet concrete in the mold may start drying, which can decrease the adhesion property.

8. Let the concrete dry thoroughly. Then remove the mold, replace the structural sill (and header, if necessary), and let down the house jacks.

A new foundation is warranted if the walls are cracked severely in most of the concrete, or if the original construction is a wood sill resting directly on the ground. In this case, again rely upon professionals.

Simple problems that ruin the appearance of concrete, as opposed to its structural stability, are something most amateurs can handle. Chipped corners and knicks and pits where aggregate has fallen out are easy patching

(A) REMOVE LOOSE PARTICLES WITH A STIFF WIRE BRUSH.
(B) WET CHIPPED AREA THOROUGHLY
(C) PROP A 2X4 AGAINST ONE SIDE OF THE CORNER. FILL CRACK WITH PATCHING CEMENT, THEN SMOOTH FLUSH WITH A TROWEL.
(D) PROP THE 2X4 AGAINST THE OTHER SIDE OF THE CORNER. USING THE 2X4 AS A GUIDE, SLICE OFF EXCESS CEMENT THAT MAY STICK OUT BEYOND THE CORNER.
(E) LEAVE 2X4 IN PLACE 4 HOURS OR MORE.

Corner concrete repair

amateurs can handle. Chipped corners and knicks and pits where aggregate has fallen out are easy patching projects that require little more than cement patch, a clean flat trowel, and squaring boards. To plumb up a battered wall:

1. Remove dirt and loose particles from the void with a stiff wire brush.
2. Wet the chipped or void area thoroughly.
3. Fill in the hole. If it is a chipped corner, prop a 2 × 4 against one side of the corner. Fill in the crack with patching cement, then smooth it flush with a trowel. Next, prop the 2 × 4 against the other side of the corner, and using the 2 × 4 as a guide, slice off any excess cement that sticks out beyond the corner.
4. Leave the 2 × 4 in place for 4 hours or more.

POST AND PIER FOUNDATION

Many older homes with common wall and footing foundations are supported by an intermediate row or rows of wooden posts on concrete piers. Because basement areas are subject to moisture and termite infestations, the wood construction here often decays before other wood areas of the house do. If any of the posts show such signs of decay as sponginess, pinholes, exaggerated raises of the grain, or extreme cracking or checking, replace them with posts treated with a wood preserva-

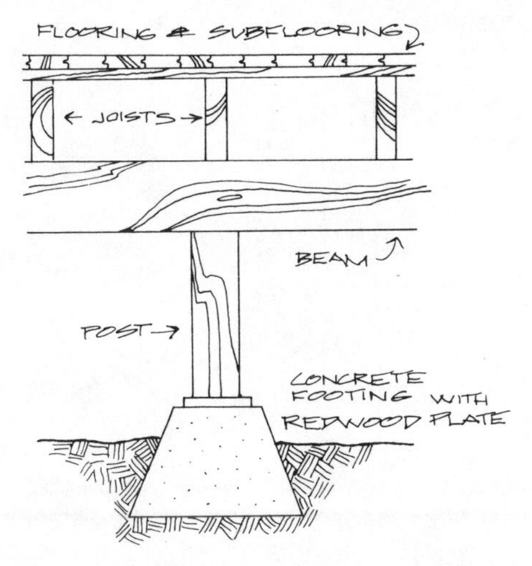

Post and pier elevations

tive. Use redwood because it is the most durable against termites. Other soft woods, such as Douglas fir, are reasonable substitutes if your budget cannot accommodate redwood. The new posts should be of the same size (larger will not hurt) as the old posts and should be placed in the same spot as the old ones.

House Jacking. To replace posts, house jack the structure. In this process the house is lifted up, just a few inches, to take the weight off of the post that is to be replaced. Consider this similar to jacking up a car to

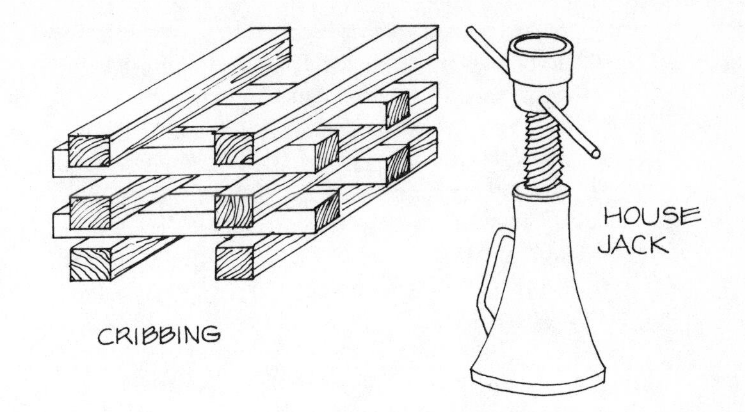

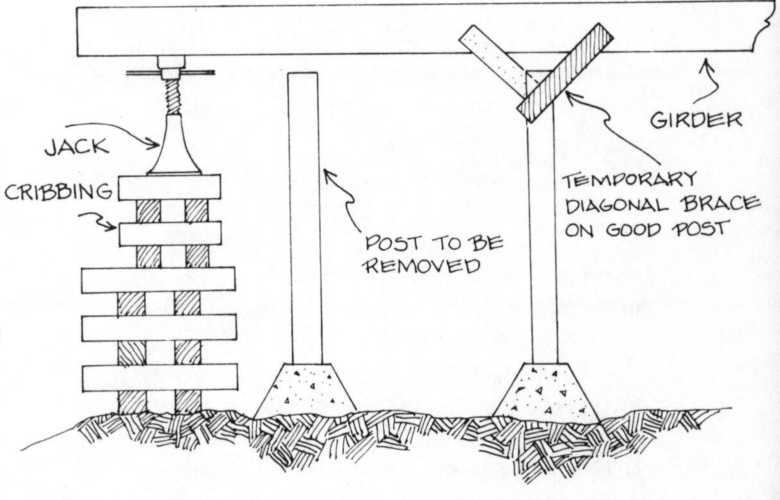

House jacking

change a flat tire. For this procedure you need house jacks, which most building supply places rent out. To lift the house, do the jacking no more than 1 foot from the post.

1. The jack (either a screw or a hydraulic jack) must be sitting on a cross-lapped web of 8 × 8-inch boards, each board about 1 to 2 feet long. This web is called cribbing.
2. Build the cribbing high enough so that the jack can almost touch the girder above.
3. Raise the jack. The house will start to creak as the jack requires more effort to turn or pump.
4. Tap or pound at the rotten post. If it feels as if it is vibrating more than the other posts, chances are the weight has been transferred from it to the jack.
5. Remove the rotten post (and pier, if necessary), and put the new support where the old one was. If the concrete pier needs replacing, be sure it is well embedded in firmly compacted soil and anchored. Center the post on the pier, keep the post vertical, and firmly anchor it to the wooden pier plate and the girder above. Hardware stores and suppliers carry steel post bases and anchoring caps. These items make the job easier and more accurate and increase the post's structural stability and resistance to lateral movement.

Wood posts exposed to the weather, like exterior porch

supports, need extra consideration when you replace them. To prevent an incubation chamber for moisture and fungus, separate the wooden post from the concrete footing with a steel elevated post base, available at hardware stores and building suppliers. This collar fits around the butt of the post and attaches to the pier, leaving a couple of inches between the wood and the concrete. This way water does not become trapped and absorbed into the end grain of the wood, and the dry rot process is greatly retarded.

SILL PLATE

The *sill plate* is a horizontal wood member that acts as an intermediary between the house frame and the masonry foundation. In some very old homes the sill plate is 8 x 8-inch denomination. In some western homes it is commonly a 3 x 4-inch, 2 x 4-inch, 2 x 6-inch, or 2 x 8-inch redwood board.

The sill should lay flat on its broadface side and be anchored tightly to the top of the foundation. Sills are in the basement area and so are highly subject to moisture and rot. You should replace decayed sills. First, remove the old sill by jacking up the house; several house jacks might be needed if the exterior wall is load bearing. Scrape off any mold, and make sure the top of the masonry wall is dry. Seal the top of the wall with a moisture repellent and a vapor barrier. (Masonry is porous and so absorbs moisture.) Get a new sill that is the same size as the old one; it should be of foundation grade redwood. Do not skimp on cost here because the money is worth it in the long run.

Bore holes in the board; the holes should directly align with the anchor bolts poking around through the top of the foundation wall. Place the sill on the wall so that the bolt ends poke up through it. Before tightening down the sill, caulk both its interior and exterior seams with silicon sealer caulking, to fill in any air gaps along the joint where the wood meets the masonry. If you saved the washers and nuts from the old bolts, use them again if you can. After the bolts are firmly tightened in place, let down the jacks, and let the new sill settle in place.

ANCHOR BOLTS

Anchor bolts are steel shanks embedded in a foundation. The bolts connect the stud wall to the masonry. They are usually ½ to ¾ inch in diameter (depending on the size of the wall and sill), 9 to 12 inches long, and available in the following three types:

1. *Common.* This is the standard bolt, hooked slightly at the end that goes inside the concrete and threaded at the portion above the wall, to accommodate a washer and nut as fasteners.

2. *Expansion bolt.* This is a type of grabbing fastener. The shank of this bolt becomes wider as it is tightened, thus providing a snugger fit within the foundation wall.

3. *No-float anchor.* This U-shaped anchor, which does not resemble types 1 and 2, can be used only where a new concrete wall is to be poured. The lower half is embedded in the foundation wall, and the upper half sticks up. The halves are bent to strap across like opposite belt ends; they are nailed to the sill plate once it is laid in place. This is an extremely useful anchor because it eliminates the need for drilling holes in the sill prior to installation, and it also lessens the chance of the sill popping up from its flush fit against the foundation wall.

Foundation expansion bolt

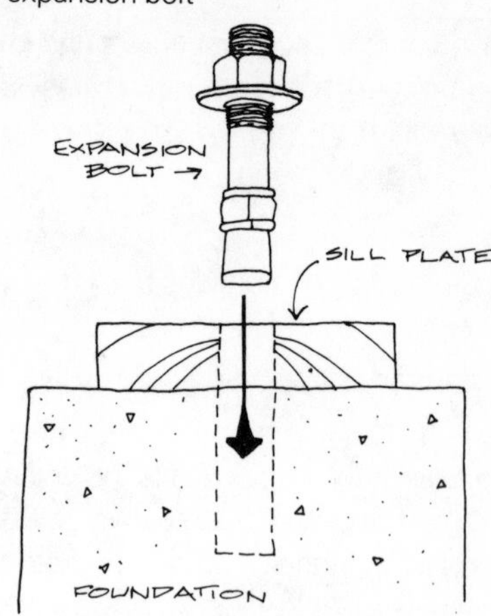

SEISMIC ANCHORS

In the course of foundation repair, consider the addition of *seismic anchors* if you live in earthquake or landslide areas. These are steel brackets or bent steel plates that bolt together the foundation, the sill, and the wall studs.

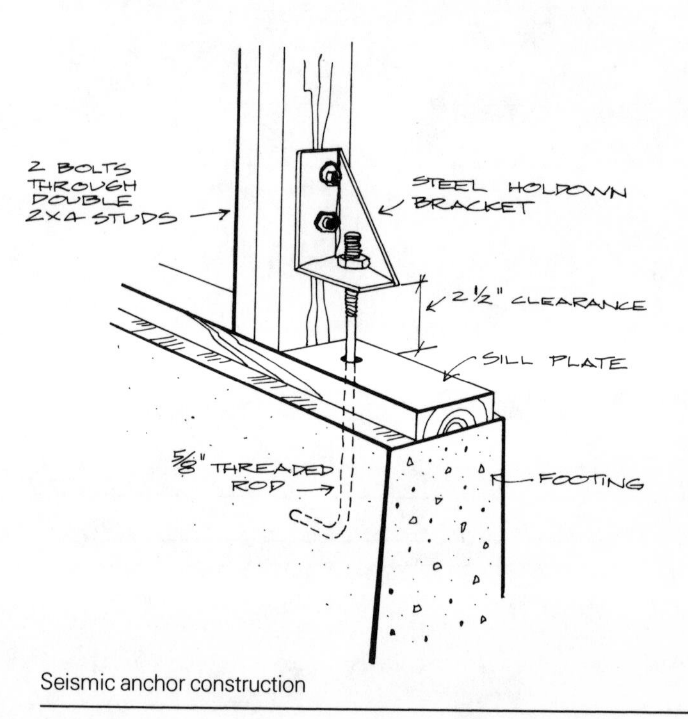

Seismic anchor construction

They should be installed at all corners of the house to protect the wood frame from excessive lateral movement. During an earthquake, or in faster-than-average settlement, the wood frame tends to become detached from the masonry foundation. The seismic anchor helps keep the unit tied together.

CRAWL SPACE

A building without an excavated basement has a crawl space, the space between the ground and the bottom of the first floor. This area is highly susceptible to moisture retention, bugs, and rodents, with moisture retention being the major cause of crawl space troubles. The moisture increases the potential for rot and decay and attracts rats and other creatures. Another discomforting characteristic is that the moist air tends to seep up into the living space above, perpetuating a rather damp and

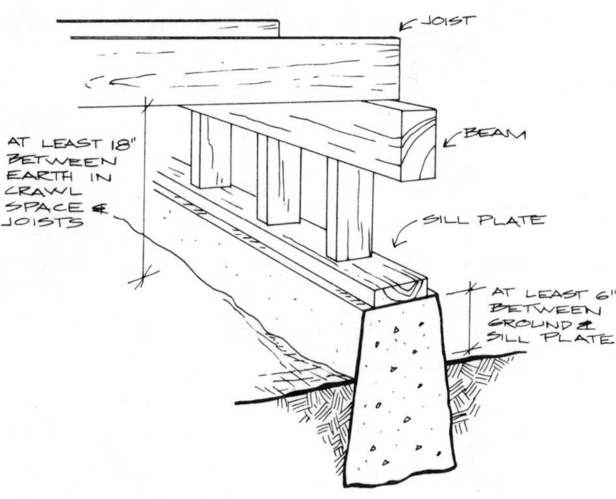

Crawl space clearance

clammy feeling in the first floor area. To alleviate as many of these problems as possible, follow these five suggestions:

1. Make sure the vertical distance between the ground and the bottom of the floor joists is no less than 18 inches. This dimension is in agreement with the uniform building code and with most local standards.

2. Treat the sills and posts with a wood preservative to prevent moisture penetration and attack from termites.

3. Keep the crawl space free from debris. Garbage or trash is an invitation to rats as well as a fire hazard.

4. Place a layer of R-11 or R-19 insulation between the exposed floor joists. Make sure the shiny vapor barrier is

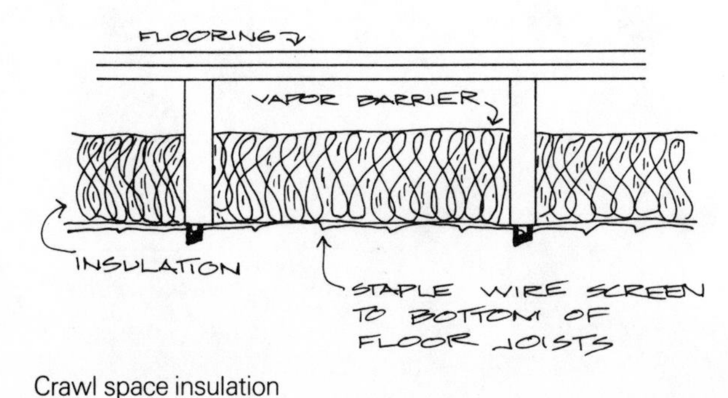

Crawl space insulation

Adding insulation

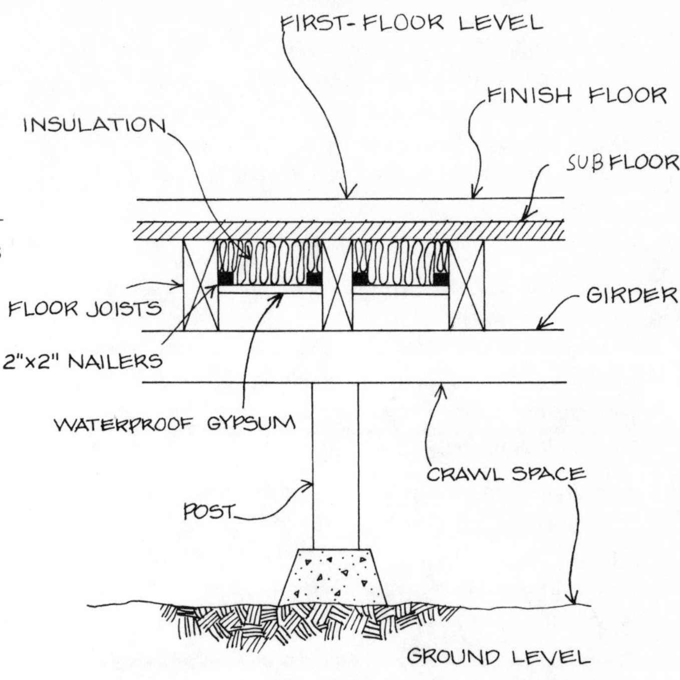

facing the ground. If desired, you can nail a covering of exterior (waterproof) gypsum board to the bottom of the joists. *Note:* The sheetrock will cover up many pipes, ducts, and wires. If you intend to engage in plumbing, mechanical, or electrical repairs, wait until this work is finished before you conceal the equipment.

5. Do not let the dirt build up against the foundation wall or concrete piers to the point that it packs against sills or posts. Maintain a 6-inch clearance between the wood and the earth. Also, make absolutely sure that no girder is touching the ground; keep girders at least 12 inches above the dirt.

Vents. Vents in a crawl space provide air circulation that inhibits moisture buildup. There should be one vent for every 150 square feet of ground area; each vent should

Vent styles

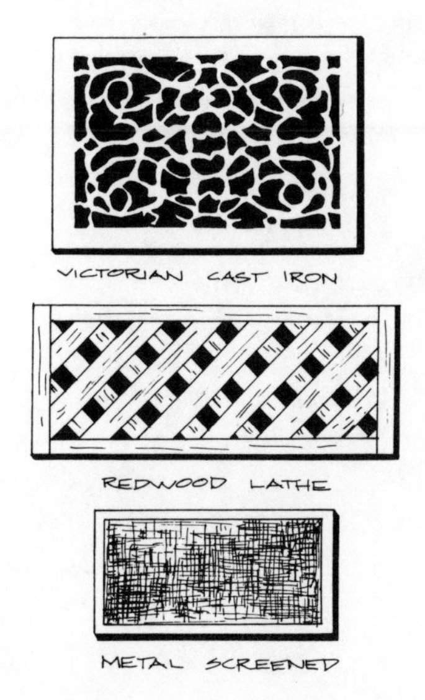

be placed opposite another to maximize cross ventilation. Keep the vents open year-round.

For greater moisture control, spread a layer of roll roofing (15-pound felt underlayment) or polyethylene sheeting (4 mils thick) across the ground in the crawl space. If you do this, you can reduce the number of vents to one per 1500 square feet. However, two vents placed opposite each other are the absolute minimum for adequate air flow.

Some Victorian homes had ornamental cast iron vents; retain the original vent whenever possible. The floral or geometric pattern of the vent allows air to enter the crawl space and at the same time decorates the building. The vent design is often masked by paint and grime, which can be stripped fairly easily:

1. Place the grate on a piece of aluminum foil whose sides are rolled up. Brush a paint remover or stripper on the vent; let it sit for several hours. The foil does not absorb the paint remover the way wood, concrete, or newspaper would.
2. When the paint remover has had a chance to do its job, hold the vent by tongs or with rubber gloves—always protect your skin from paint remover—and hose or scrape off the gummy surface.
3. If paint still adheres, repeat the process.
4. Use steel wool to rub off the final specks.

Now that the crispness of detail has been restored, the vent is ready for a protective primer and a fresh coat of paint.

Shingle Victorians and Classic Bungalows had vents made of crisscrossed redwood lath. Unfortunately, this design is susceptible to rot and may allow rodents and other pests to enter the crawl space. Without sacrificing style, you can insert a metal substitute. But if you install a new redwood lath, thoroughly treat it with a wood preservative.

Many houses with wood siding at ground level have holes drilled through the siding in lieu of a more conventional vent. Air flow may be insufficient, and these holes can also serve as an entry for yellow jackets. Many buildings from the 1920s have galvanized mesh metal vents, which usually rust, bow, and tear after several years and so require replacement.

Although many of the old vents are more decoratively distinct than the contemporary ones, their opening is large enough to let rodents and insects enter. Vent openings should be no larger than ¼ inch square to function successfully. To save an interesting vent that exceeds this standard, attach a screen to the inside of the vent. If the vent is neither operable nor noteworthy, use just a screen vent.

[PART TWO]

EXTERIOR ELEMENTS

The exterior of your old house is the ornamentation and decoration, line and design—it is in total what constitutes the Victorian. This, as has been mentioned, may be a combination of various periods—Colonial or Period Revival, Classical, Gothic, and so on. In any case it is the exterior and its embellishments that say Victorian.

The exterior of the house is the basic sheathing, the roof, the windows, front door, stairs, and porch. It is the face of the house, and as human faces age with time, so do houses. There is always work to be done on the facade—the basic outside—but restoring the ornamentation is vital, as is putting back into good condition the basic materials that constitute the shell of the home.

[CHAPTER 4]

ROOFS

A leakproof roof is essential to the longevity of the entire house. If water enters the building, it will decay the wood structure and damage exterior siding and interior finishes. Roof repair is therefore one of the first steps in restoration. Examine all the roof, inside and out, from its weather-exposed finish to the structural rafters that support it, because every component will determine the ultimate extent to which you must consider repair work.

[OUTSIDE DAMAGE]

The roof is the cap on the house's head and definitely should be considered very much a part of its overall appearance. But too often this area is neglected aesthetically. People tend to regard the roof as merely a functional necessity and end up crowning a worthy face with thorns. Whether highly visible or more subtle in nature, the roof *can* be seen, if not by the house's occupants, by the neighbors. Just as the ground row of shrubbery defines the house in relation to the earth, the roof provides the finishing serif of structural expression.

The first step you absolutely *must* take to diagnose your roof is to climb a ladder and get on top of the roof. Other than by noting water stains on the ceiling of a room, you simply cannot tell what is going on above your head without elevating yourself above eye level. Once on the roof, check for:

1. Missing, broken, or worn-out shingles or tiles in a pitched roof;

2. Ruffles, separations, or cracking in the asphalt or roofing felt on a flat roof;

3. Loose flashing, especially around the chimney and valley;

4. Sagging ridges;

5. Protruding nail heads.

Minor leaks in the roof can be repaired with an asphalt roofing compound and a liquid roof coating. However, every roofing material has a definite lifespan, so when the roof starts to leak because of general material failure, it is time to install new roofing.

MATERIALS AND DESIGN

When you select new roofing material, take into account visual and functional considerations. Functionally, the questions of durability, fire resistance, slope and cost are important. (Table 5.1 compares these qualities.) Fire resistance is especially important because many local ordinances, particularly in urban areas, prohibit the use of combustible materials as roof finishes.

Visually, the selection relates to the shape of the roof, the amount of roof surface seen from eye level, and the material appropriate to the architectural style. Sometimes you cannot completely re-create the original roof characteristics because present laws do not permit certain materials. Nevertheless, the flavor and essence can be preserved by compromising certain colors, textures, and patterns of other materials. For instance, if you have to replace a thoroughly deteriorated wood shingle roof, but local ordinance does not permit the use of this combustible material, use a brown asphalt shingle, the color and scale are such that it will relate to the original wood shingle.

Providing the proper roof can sometimes be a ticklish problem because many roofs have been relayered so many times that it can be difficult to determine exactly which design and material is historically correct. Most likely the Victorian, Brown Shingle, Colonial Revival, and Classic will have their original roofs of wood shingles or shakes. Mediterranean-style (stucco) houses invariably have roofs of barrel or other ceramic tiles. These tiles will probably still be intact because they are noncombustible and not prone to rapid decay the way wood tiles are. Slate and copper roofs are not that common on West Coast residences. If you *do* have an original slate or copper roof, guard it with your life. These are not only highly unique materials but extremely valuable.

When reroofing, do not drastically change from the previous material. If the original roof is tile, for instance, repair it with the like; do not tear off the whole roof just because you prefer wood shakes. Likewise, never reroof with tile a house that calls for wood shingles. Choose only wood shingles, or, if this material is prohibited in your area, consider an earth- or wood-colored asphalt shingle. Gray asphalt shingles are also a decent solution because the clean neutrality of gray places greater emphasis on the form of the roof.

Asphalt shingles have been often misused (people have applied the multicolored, 1955 types) and so they have gained a reputation of being cheap, flimsy, and artificial looking. However, the asphalt shingle roof is fine if you use colors reflective of the house's period. This material is a good second choice if you need to reroof but are on a limited budget.

FLASHING

Flashing is a strip of thin-gauge sheet metal that prevents water from entering the building. It covers the exposed joints and diverts the water to less vulnerable areas. Flashing is found at:

1. Roof valleys, ridges, hips, and changes in pitch;
2. Door and window openings;
3. Expansion joints;
4. The juncture of the building and porch;
5. Vertical projections through the roof, like skylights, vent pipes, the chimney, and dormers;
6. The juncture of the building and ground.

Modern flashing is typically galvanized steel. On older buildings the flashing was either copper, lead, or a zinc alloy; it is often badly deteriorated, rusted through, or

TABLE 5.1: COMPARISON OF ROOFING MATERIALS

Material (in order of durability)	*Minimum Slope (in in.)*	*Approximate Roofing Cost for 1500 sq. ft.[1] (cost per sq. ft. in parenthesis)*	*Fire Rating*
Slate	4″ in 12″	$10,000 ($6.67)	NC[2]
Copper	½″ in 12″	6,000 (4.40)	NC
Ceramic tile	4″ in 12″	.2,000 (1.44)	NC
Tar and gravel	Flat	660 (0.44)	NC
Asbestos tile	5″ in 12″	1,900 (1.27)	NC
Wood shakes	3 to 6″ in 12″	1,300 (0.87)	C[3]
Wood shingles	3 to 6″ in 12″	1,300 (0.87) [if fire rated: 2,300 (1.53)]	C
Galvanized steel	3″ in 12″	1,400 (0.93)	NC
Asphalt (composition shingles)	2 to 3″ in 12″	600 (0.40)	NC
Roll roofing	2 to 3″ in 12″	600 (0.40)	NC

1. These are very approximate costs and vary greatly, depending on the number of different roof surfaces, how they intersect, how accessible the roof is to workers, and the ever-changing prices of materials and labor. Use the table to compare the relative cost of one material with the other, not as an accurate guide to roofing costs for your house.

2. Noncombustible 3. Combustible

Porch roof flashing

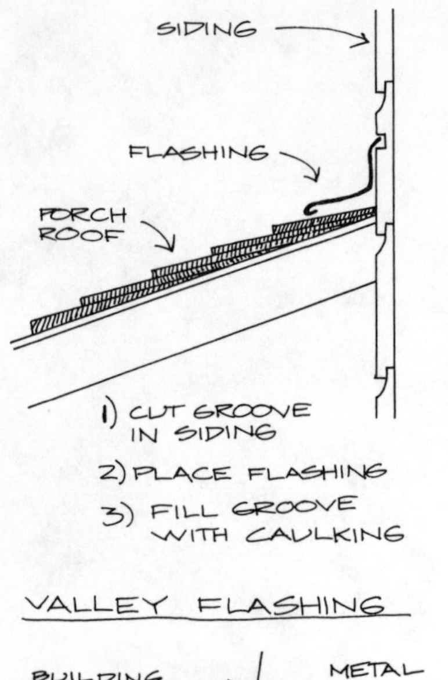

VALLEY FLASHING

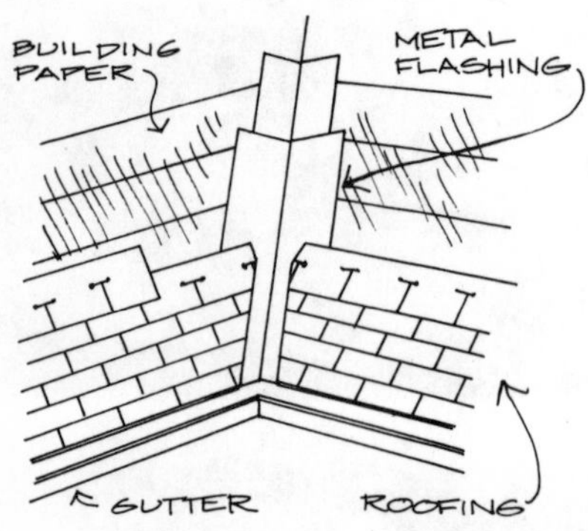

missing altogether. This is an obvious source of serious leakage and should be corrected early in the restoration process.

At vertical projections, two pieces of flashing are required: the base flashing, which keeps water away from the joint; and a cap or counterflashing, which keeps water from sneaking behind the base flashing. Only one piece of flashing is required for the valley, ridge, eave, window, and door locations.

To repair small holes in flashing, for each hole cut a pieces of sheet metal 1 inch larger on all sides than the hole. Coat the hole and the immediate area with roof cement. Press the sheet metal to the hole for several minutes, until the material adheres to the cement.

If the holes are widespread, or the flashing is generally deteriorated, replace it. This is no major job, entailing little, if any, alteration to the structure. Just nail the strips in place, making sure the flashing has no warps and kinks that might divert the water into undesired areas. When installing flashing, be sure that all the nails and clips are of the same metal as that of the flashing (or compatible with it). When dissimilar metals are in contact with each other, a chemical reaction occurs, which causes the deterioration of one of the two metals. For galvanized steel flashing, use zinc-coated fasteners.

On a house with a parapet wall, be sure the flashing extends at least 12 inches above the parapet because this area is very susceptible to water accumulation from clogged gutters. On a house with a front porch roof, be sure there is adequate flashing to protect the vulnerable joint between the porch roof and house siding.

GUTTERS AND DOWNSPOUTS

Gutters collect water that drains off the roof and direct it to the downspouts. Some Victorian homes had decorative redwood gutters (sometimes resembling classic stone moldings) that were intended as part of the ornamentation. If the redwood gutter has rotted, but the decay is limited, you can repair it. With a sharp tool, scrape out the rotted material until you reach solid wood; then fill the scar with wood putty. If the rot is more serious, but localized, remove the bad segments and splice in matching new pieces. Localized rot is common around downspouts or at joints between gutter sections. If the rot is widespread, put in a new gutter. Some specialized places do carry redwood gutters. Before installing them, treat them with a sealer/preservative to retard the wood from decay.

Paint wood gutters every two years or so. On the inside of the gutters, use two coats of asphalt roof paint thinned to brushing consistency (one part thinner to four parts paint). On the outside, apply two coats of an appropriate house paint.

Galvanized metal gutters are sold at almost any building-supply store; they usually cost less than $3 for a 10-foot section 4 inches wide. To reduce the offensive glare of the metal and help it harmonize with the architectural style, paint the gutter with a neutral matte finish or the color of the house trim. You can also buy rust-free aluminum gutters that are prefinished with baked-on enamel.

Any gutter you select should reflect and relate to the original gutter design in size and cross section for better water control and authentic appearance.

Queen Anne and other turreted house styles have curved gutters, which are difficult to replace. This is why you often see such houses with only partial gutters. If you are willing and able to pay the price, you can have curved gutters custom made out of sheet metal for about $7 a lineal foot by a gutter/downspout contractor.

Some Victorian homes, such as the Brown Shingle, were built without gutters. As a result, water accumulation often caused soil erosion below the roof overhang and premature blistering of the paint on the wall that faced the prevailing wind. If you have these problems, consider adding gutters that relate to the architectural style of the house.

Check downspouts for leaks, especially at joints, at the gutter connection, and anywhere the downspout is not vertical. If the downspout empties directly onto the

N
S

Good and bad downspout locations

THE HOUSE ON THE LEFT
HAS A DOWNSPOUT LOCATED ON THE SIDE, FAR ENOUGH BACK TO AVOID INTERFERENCE WITH THE PORTICO. THIS IS A GOOD SOLUTION.
THE HOUSE ON THE RIGHT
HAS TWO DOWNSPOUT PROBLEMS. A DOWNSPOUT INTERRUPTS THE VISUALLY IMPORTANT STREET FACADE, DIVIDING IT IN TWO. ALSO, IT IS PAINTED A CONTRASTING COLOR, EMPHASIZING THE DIVISION.

ground, place splashblocks at the outflow to break the fall of the water and minimize erosion around the foundation. Use brick, stone, or a precast concrete product, and slope it away from the house. Never connect the downspout to sewer hoppers or sanitary sewer systems.

If you are adding a downspout, locate it as inconspicuously as possible, and paint it the same color as the wall behind it; the corner, side, and back of the building offer much better camouflage than the face.

EAVE, FASCIA, AND SOFFIT

The eave, fascia, and soffit, all found at the upper perimeter of the roof structure, are valued for their ornamental properties. Made of wood, these elements are subject to rot from leaking roofs, faulty gutters, poor design, and deferred maintenance. For example, if the gutters become clogged with leaves and water cannot drain, the water overflow will soak the fascia. If the situation persists, the fascia, soffit moldings, and rafter ends may decay.

Water tends to penetrate the eaves at the edge of the roofing. To prevent this from occurring, install a metal drip strip between the roof sheathing and the roofing material, as illustrated.

The gutterless Brown Shingle houses and Classic houses have rafters that extend well beyond the roof and are normally left unpainted. Water rolling off the gutterless roof onto the exposed rafters encourages rot; the end grain is the most susceptible spot. Inspect the end grain of the beam to evaluate the extent of damage (wood feels spongy, shows splits, or is charred). If damage is very limited, remove the tip of the rafter. If rot has permeated the whole piece, graft on a new section, as discussed in the illustration for stickwork repair. Be sure to protect the end grain of all rafters, new or old, with a clear sealer.

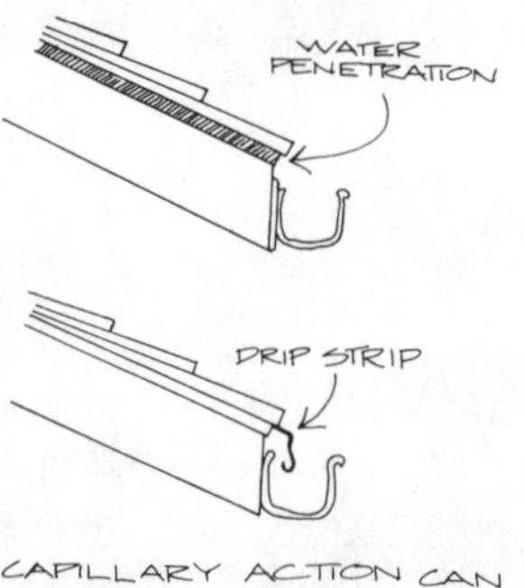

WATER PENETRATION
Drip strip
CAPILLARY ACTION CAN DRAW WATER BETWEEN SHINGLES & SHEATHING, CAUSING DAMAGE TO BOTH. INSERTION OF A DRIP STRIP PREVENTS THIS WATER DAMAGE.

[DETECTING ROOF DAMAGE FROM INSIDE THE ATTIC]

Many telltale signs of potential threat or existing roof damage can be detected from inside the attic. Look for rays of sunlight or water stains on the rafters or roof decking, obvious signs that the roof is not watertight. If the wood inside is not rotten but merely stained, scrape and clean away any mold or mildew, and treat the wood with a sealer. Then patch the necessary areas of the roof. Repair minor leaks with asphalt roofing compound and liquid roof coating.

If the roof rafters are rotten or sagging to an uncomfortable degree, small patch jobs are insufficient. Correcting this unfortunate situation entails removing the entire roof surface and replacing the rotted rafters. Consult a roofing contractor to determine the specific problems and remedies.

If you elect to undertake some or all of the repair work, make sure you do so during fair weather because you may possibly run into corrections that will take longer than you had anticipated. If you are literally rebuilding

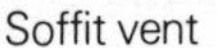
Soffit vent

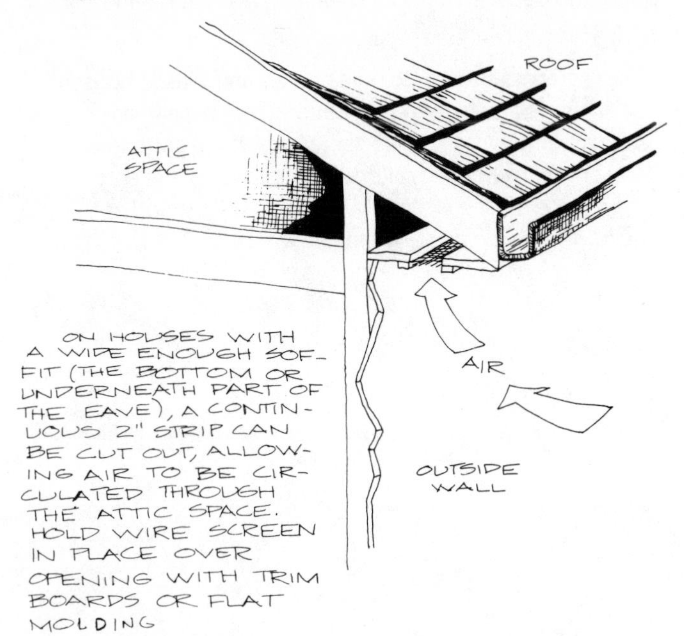

the roof, be sure that all the rafters are of the proper dimensions and angled into the identical slope as the originals. Do not rely upon the original framework to dictate the lumber sizes to use because it may be substandard in relating to present building code. First check all the sizes, insulation, spacing, and decking requirements, which differ from roof pitch to roof pitch, style to style, and area to area.

ROOF
ATTIC SPACE
AIR
OUTSIDE WALL
ON HOUSES WITH A WIDE ENOUGH SOFFIT (THE BOTTOM OR UNDERNEATH PART OF THE EAVE), A CONTINUOUS 2" STRIP CAN BE CUT OUT, ALLOWING AIR TO BE CIRCULATED THROUGH THE ATTIC SPACE. HOLD WIRE SCREEN IN PLACE OVER OPENING WITH TRIM BOARDS OR FLAT MOLDING

VENTS

Moist, warm air rises and then condenses in the cooler attic space. The attic must be properly ventilated to allow moisture to evaporate and to prevent rot and structural damage to the rafters. Locate and reactivate the original vents. The net area for the openings should be about 3 percent of the ceiling area of the room below. If the original vents are too small or not arranged to permit adequate air circulation, you can add vents. On house styles with an overhang—such as Queen Anne, Colonial Revival, and Period Revival—vents under the soffit are easy to install, leak resistant, and well camouflaged.

For a house with insufficient overhangs and inadequate air ventilation, consider installing a roof-mounted ventilator. This is a light metal duct with a cap on top; it acts like a fan when blown by the wind, sucking air out of the attic. Prior to cutting the hole for installation, make sure the roofing is in good enough condition to be sealed up again. Locate the roof ventilator so it will not be noticeable from the existing vent to ensure air circulation. Paint the vent a neutral, matte-finish color.

Cutting new vent holes in a major part of the house's facade may ruin the visual authenticity if you are not careful. Consider the style of the house, and then locate the vent so it blends in with the other architectural features.

(CHAPTER 5)

EXTERIOR SHEATHING

Sheathing (siding) is the weatherproof surface applied to the completed structure of the house. It protects the house's structural components, including the plumbing, wiring, and fixtures, from the elements. Any air draft passages force the heating system to work harder, and cracks let in moisture that seriously damages the interior.

Visually, the original sheathing complements the other architectural features of the house. The design detail, color, and proportional breakup of the siding are what give the house its impression. You have probably noticed a house that had just had its planking replaced with shingles. Most likely your response was "It looks like a different house," or "I almost didn't recognize it." The sheathing can make a world of difference in the entire appearance of a house.

When considering repairing the sheathing or replacing it, you should try to match the original siding if it was wood. However, sometimes the original siding is difficult to detect. To meet fire codes, many wood houses have been refinished with stucco, aluminum siding, or asphalt or asbestos shingles. If a wood siding is still intact but needs some patching or replacing, attempt a face lift that will re-create the original features.

If your Victorian is brick, it probably has not suffered much weather damage, nor has anyone attempted to resurface it with another type of sheathing. (The California Mission and Spanish Revival styles were originally finished with stucco, so by no means replace the siding with wood.) Whatever the sheathing is, always repair it in kind. Trying to save work by changing to modern, mass-produced or synthetic sidings defeats the architectural integrity and lowers resale value.

(STUCCO)

Stucco is a composition of Portland cement, sand, and lime. It is applied like plaster to the exterior of wood frame houses. The use of stucco as siding gained popularity throughout the United States during the 1940s and 1960s. Historically, the styles to which stucco belongs are some California Classic Bungalows, the Spanish Revival, and Mission. Unfortunately, stucco is misused in attempts to "modernize" Victorian houses whose rightful weatherproofing is drop siding, bevel siding, or shingles. If original wood siding is buried underneath stucco, definitely remove the stucco if local codes permit. (Stucco was once the only solution to regulations that require a 1-hour fire-rated exterior wall on any hotel or

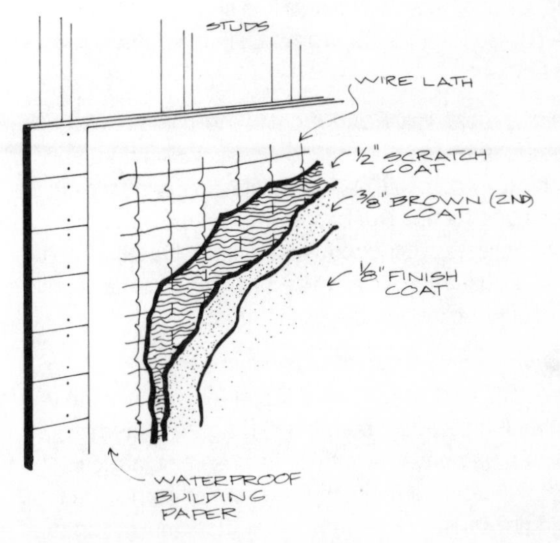

Stucco construction

apartment house three or more stories tall. Thus, many three-story single-family Colonial Revival and Brown Shingle houses that were converted to multifamily dwellings now have stucco siding.)

REMOVAL

Attack the stucco a section at a time with a hammer and chisel. Make a 2-inch-wide square-shaped score about 3 feet long on each side. Pry off the stucco with a crow bar. Use clippers to remove the chicken wire, and your hands to rip off the building paper. Remove the nails with a claw hammer. The wood siding will be in remarkably good shape if it has been kept dry and unexposed to the elements. Before repainting the wood surface, be sure it is free of dirt and all the holes and gouges have been patched.

REPAIR

Stucco is prone to surface damage, which begins with hairline cracks, usually near windows and doors, and deeper damage, which begins with leaks in the flashing or inadequate ventilation in the crawl space. Repairing narrow surface cracks to prevent more serious damage is fairly simple to do:

1. With a knife or a spatula, open the crack until you reach sound stucco.
2. Use a hammer and cold chisel to make the edge of the crack wider on the inside than at the outside edge. This inverted V will lock in the new stucco. Brush away all loose material.
3. Prepare a dry-mix mortar, adding water until it has a firm yet pliable consistency. Dampen the crack and pack the stucco in tightly with putty knife or trowel. Overfill the crack if it extends through the stucco to the base material. Let the mortar dry for about 15 minutes, and then work it down until it is flush.
4. Moist cure the stucco with a fine spray from the garden hose for about 3 days, once in the morning and once at night.

Repairing major patches is more complicated because the procedures require attention to the wood framing, sheathing, wire mesh, and the coats of stucco (see earlier stucco construction illustration). Stucco reconstruction is really a delicate art and requires professional skill. The time and frustration that can be saved by hiring a stucco specialist is well worth the expenditure.

If you want to repair small areas that have been nicked, chipped off, or damaged in some minor way, be sure that the sheathing is intact before replacing the waterproof building paper and wire lath. If the sheathing appears free of moisture damage, the patching of a small area should not cause any trouble, other than duplicating the original finish. The last coat of the new patch must match both the color and texture of the rest of the wall. Provide a sandy finish by trowelling the stucco with a float as the stucco begins to set; for a swirled finish, use an old brush to scrub or jab the surface.

Stucco has a natural color all its own, originally appreciated for the graininess the sand and cement imparted without benefit of paint. If the house is still this distinctive, unadulterated gravelly gray, leave it as it is.

If the stucco mixture was originally treated with a color, do the same as you patch it. Add mineral pigments to the mortar for the final coat, mixing the mortar well enough to produce a uniform color. On large jobs, the pigment should weigh no more than 5 percent of the

Types of wood siding

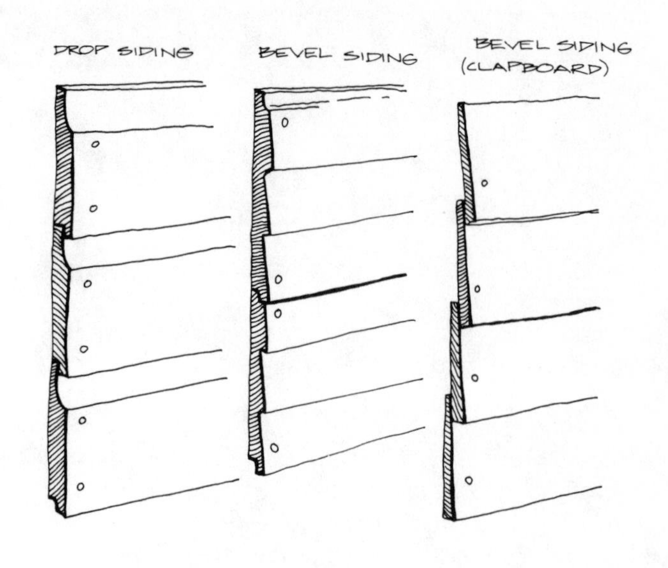

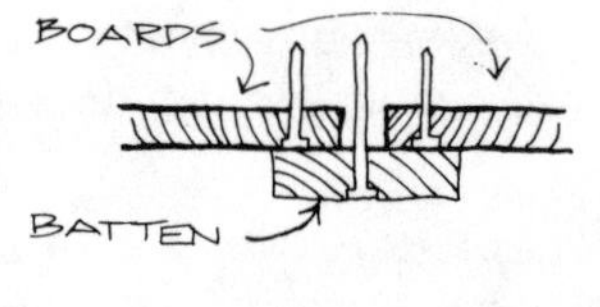

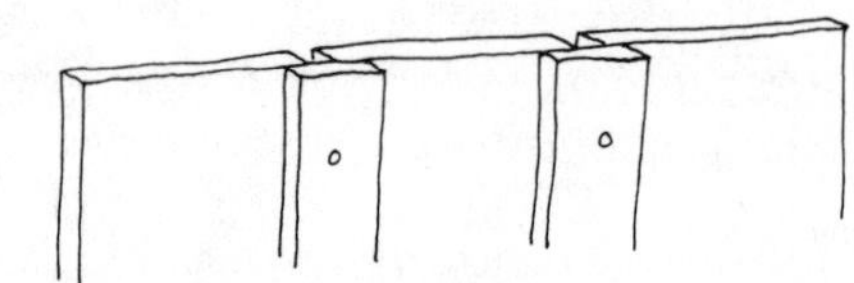

total weight of the masonry cement. This is difficult to calculate in small quantities, so use a cook's teaspoon to add a little bit of the powder at a time. Note your final recipe right on the box of pigment for future reference. Available stucco colors include pink, yellow, green, and tan.

If the stucco has a decidedly painted surface, repaint the house. (See Chapter 9 for paint-selection instructions.) If you wish to paint a previously unpainted stucco surface, be sure to first coat the walls with a sealer or otherwise the paint will be excessively absorbed, producing an inconsistent and blotchy surface.

[WOOD SIDING]

Wood siding was the most widely used residential weatherproofing material for Victorians. Except for Classic Bungalows, whose finish should always be left natural, always paint wood siding.

DROP SIDING

Most prevalent in Victorian era homes was a horizontal sheathing called *drop siding*; this is characteristized by flat, interlocking panels whose joints have either rounded channels or V-shaped grooves. This design was also occasionally used as interior base molding. Original drop siding planks were a good 6 to 10 inches wide, with a distinctive ½-inch dividing channel between the planks.

BEVEL SIDING

Bevel siding is slightly angulated to the vertical. Each horizontal plank is tapered at the top, wider at the bottom. The simplest kind of bevel siding is *lap,* or *clapboard*; each plank is nailed to overlap the plank below it. The width of a clapboard plank is usually 8 inches. Another type of bevel siding employs the characteristics of both drop and clapboard style; the joinery between the boards is notched or rabbeted as in drop siding. This type of construction joint creates a virtually watertight seal.

A third type is *shiplap* siding, which is flat and flush; it looks like tongue and groove flooring, and it is installed either horizontally or vertically, a versatile quality not effectively possible with drop or bevel siding. Single planks are divided into three beveled tiers, with the rabbet joint at every set of bevels. Visually this creates a more pronounced horizontal line in every three slats, which subtly but very effectively moderates an otherwise repetitive design. Shiplap joinery in bevel siding is found on many Colonial Revival homes.

BOARD AND BATTEN

A common wall sheathing, *board and batten* is a vertical siding of standard lumber nailed to a substrate of plywood or tongue and groove; the boards are either side butted together or placed with a ½-to 1-inch space between them. A less wide piece of wood, the batten, is face nailed over the seam, to create a vertical corrugated surface. Like clapboard siding, board sizes vary: 6-, 8-, and 10-inch exposures are the most common.

REPAIR

Cracked or decayed wood siding is an invitation to further trouble, especially from water seepage. To determine if there is damage, pick and pry at various planks to see if they are soft and brittle rather than hard and resilient. If the boards are firm and solid, they are undoubtedly rotten or have been munched on by animals. Also check the interior sides of the walls; peeling paint and bubbling up or discolored interior wall surfaces are signs that moisture has taken its toll. This entails rebuilding the framework and resurfacing the interior as well as replacing the rotten exterior siding.

You can fill small fissures in the siding with putty or caulk. Scrape the surface clean of any old paint; clean the crack of any paint or dirt. Smooth away projecting splinters, nicks, and the like. Squeeze the caulking into place and let it dry. For cracks that are too wide to be sealed with caulk alone, first stuff them with *oakum*, a specially treated ropelike material available on spools at plumbing-supply stores.

Major cracks and deterioration require replacement boards. If an entire board must go, remove the nails with a nail puller (which can be rented for several dollars a day); countersinking of the nails and layers of paint over the years usually make the use of a hammer claw alone unfeasible. As needed, replace rotted building paper, and repair punctures with asphalt roofing compound. Treat the new wood with a *penta* (pentachlorophenol) preservative that can be painted on. Slip the new slat in place, and secure it with aluminum or galvanized nails. Countersink the nails, putty the holes (including those made from the nail puller, caulk the joints, and paint.

If less than half a board is damaged, it is still wiser to replace the entire piece because partial replacement entails time-consuming, skillful operations: You have to saw off the damaged segment from the wall, which involves careful manipulation of a circular saw. You may damage the adjacent siding and underlying building paper.

The most pressing problem when replacing siding is matching the original. If your siding is clapboard or clapboardlike, you need not be too concerned, but finding replicas of drop siding with round or V-shaped edges may be a headache. Victorian restoration suppliers may carry what you need or be able to direct you to where you can obtain it. If not, you can try your skill with a router. First obtain a milled (redwood only) board. Examine the cuts made in the original plank, and use the proper router bits to carve the identical pattern. It is usually not too difficult to choose the correct bits, because a pictorial description of the cut made by the bit is usually included on the package.

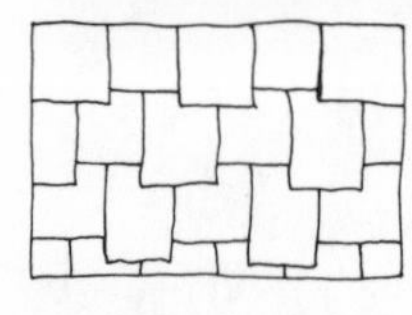

SQUARE BUTT

OCTAGON

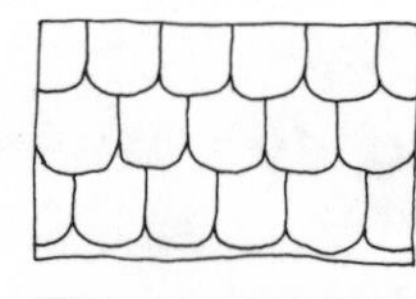

FISHSCALE

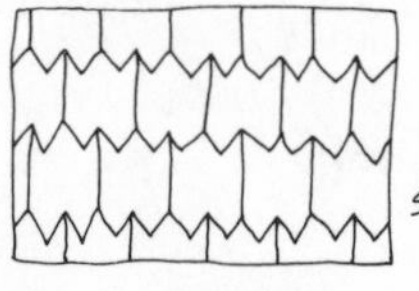

SAWTOOTH

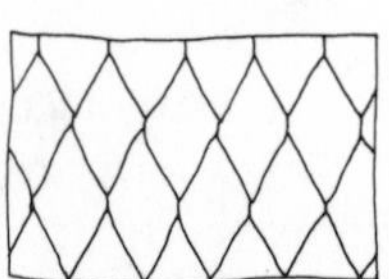

DIAMOND

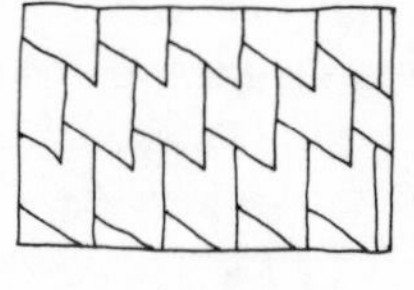

CHISEL

If your ailing siding is board and batten, you are lucky because this is much easier to repair than any horizontal siding, since the boards and battens are face nailed and will accept pry bars.

[WOOD SHINGLES]

Wood shingles are a flexible siding material because they can fit around obstructions and cover misshapen walls. The standard shingle is 16-, 18-, or 24-inches long, made of redwood or cedar. Shingle siding is a characteristic of Californian Classic Bungalows and Brown Shingle houses. Shingles come in various shapes, including rectangular, octagonal, diamond, fishscale, and others (see illustration). The Queen Anne Victorian is noted for its quiltlike, geometrical shingle patterns, as are certain Provincial homes sided with heavy, coarse, irregularly shaped shakes installed with 1 to 2 feet exposed to the weather. However, shakes, popular in today's "hand-built home revival" look, are really inappropriate as siding material; they are functionally and aesthetically better suited as roofing because their heaviness withstands sun, rain, snow, and hail. Shakes overpower style details and look ugly. Never replace walls originally shingled with shakes.

REPAIR

Shingles are sold in packages called squares. A square is made up of four bundles; each bundle covers about 25 square feet. There are two different grades of cedar shingles, No. 1 and No. 2. The No. 1 grade has a more consistent grainage, and it splits more easily and in a straighter line than does a No. 2 grade. The No. 1 grade costs 20 to 25 percent more than the No. 2 grade. Custom-shaped shingles, like those illustrated, are usually only available at historical suppliers and cost about twice as much per quantity as the conventional No. 1 grade shingle.

Replacing outworn shingles is a good job for an amateur. Soak a bundle in a bucket of water several hours before applying the shingles. This prevents the shingles from swelling during the first rain and possibly popping. To remove a damaged shingle, slip a hacksaw blade under the bottom of the good shingle above it and cut the nails that hold the top of the damaged piece. Similarly, if necessary, cut the nails at the bottom of the damaged shingle. Splinter the shingle with a chisel, and pry out the nail stubs with the claw of a hammer or with pincers. Slip Slip a new shingle of the same size, thickness, and approximate color into the space. The top of the new shingle should be overlapped by the upper shingle course; the bottom of the replacement should overlap the lower shingle course.

Shingles can be finished with stain, preservative, or paint. Painted shingles are appropriate to the Queen Anne style. On a First Bay Tradition house (San Francisco Bay Area), paint is not desirable because it is architecturally out of place and clots the pores of the wood, preventing the grain from expiring any absorbed moisture. Sometimes people paint shingles to conceal discoloration, but this darkening often bleeds right through the paint. It is better to apply a fresh coat of stain.

If the shingles on a Classic Bungalow or Brown Shingle house have been painted, and you want to repaint, the most authentic color to use is a raisin brown, which simulates the appearance of naturally aged redwood.

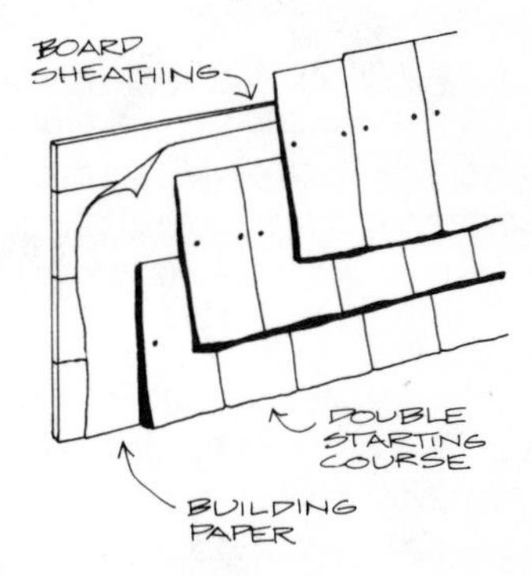

PLACE NAILS ABOVE TOP OF PREVIOUS COURSE & 3/4" FROM EACH EDGE, 2 NAILS PER SHINGLE. NO MORE THAN 7 1/2" OF A 16" SHINGLE SHOULD BE EXPOSED.
BOARD SHEATHING
DOUBLE STARTING COURSE
BUILDING PAPER

Shingle installation detail

The brown, unlike the pea-green or gray the house was most likely painted, provides a visual foil for the rich green of nearby shrubs and trees.

If the shingles are still natural and have acquired an enviable patina with age, maintain them with a coat of preservative. (Cedar turns a silver gray, redwood a raisin brown.) The difficulty in replacing natural shingles in scattered locations is matching the new slab with the weathered look of the rest of the wall, thus avoiding a buckshot look. If you anticipate repairs in the next year or so, buy a square of shingles now and leave them in a safe spot outdoors to weather.

If the shingles have been stained, stay with stain. Paint's opague pigment hides the wood beneath it, but stain has enough pigment to just color the wood without concealing the grain or suffocating the fibers. Check the label or ask a paint dealer for a stain that contains a preservative.

To match new stained shingles to the original ones, add drops of stain to paint thinner until it tests a bit lighter than the original shingles. Brush the solution on the new shingles in the same direction as the grain. The pigment darkens the wood enough to camouflage its youth, and the thinner reduces the oil content of the stain, which would otherwise deter natural weathering. Use the same staining technique to make new cedar shingles blend

Shingle patterns

with redwood originals (redwood shingles are practically impossible to find). When selecting any stain, remember that the color looks darker in the can or on a swatch than it does in natural light on a large wall.

[MASS-PRODUCED SIDING]

Beginning in the 1950s, mass marketing experts conspired to alter the appearance of the nation's homes by manufacturing and promoting mass-products siding that would "modernize" a house. Because these items were not designed for any particular architectural style, they are so anonymous that they go with no style at all. They were conceived to look enticing in a magazine ad or in a display case, but when applied en masse, the appearance is tawdry, flimsy, and just plain imitative. Other items, available by catalog number and off the shelf, instead of by design characteristics became the mainstay of the new pastime called "home improvement." Most of these products — flush doors, aluminum frame windows, plastic ornamentation — were not too bad on modern homes, but they are not authentic for Victorian restoration.

One of the greatest crimes against a historically significant building is to apply the tawdry makeup of mass-produced synthetic siding. It is ludicrous that certain types of this siding pretend to be something they are not: plastic called "stone," asbestos called "shingles," aluminum shaped like clapboard. The imitation is never convincing, and the superficial aspect is blatantly obvious.

Asbestos "shingles" are brittle, thin, corrugated, tilelike slabs made of mineral fibers. They are 12 inches high and broader than the wood shingles they try to imitate. Asbestos shingles come in nondescript gray-green, gray-pink, and speckled tones. Aluminum and vinyl siding are extruded pieces of metal and plastic, respectively, shaped like shiplap but much thinner and lighter in weight. Aluminum tends to dent, is noisy in the rain, and peels off if not properly anodized. Vinyl reacts with light in the atmosphere, becoming brittle and deteriorated if it is not treated with an ultraviolet inhibitor.

Tarpaper brick is like an asphalt roofing material. The "grout" is represented by the gray background, the "bricks" by rectangles of tan or overly bright red with the texture of sandpaper. Other disastrous substitutes include ground stone or brick crumbs suspended in plastic. A fibrous spray that sheathes the building in a filmy cloud is, thankfully, a little-used atrocity. When used as siding, these products are sure-fire ways to create an overpowering sense of aesthetic destruction.

Aside from the obvious injustice modern siding does to an old house's architecture, it has other disadvantages. For example, there is the hidden expense of adjusting window casing, drip cap, and door trim to compensate for the added wall thickness. Visually, the modern siding replaces "line" with "texture" and makes the building look choppy instead of solid. Breaking up the facade into different surfaces also reduces the impact of the house. A variety of textures or materials presents the building as a conglomerate of detail, as opposed to one integrated statement. The so-called random pattern that artificial brick and stone sidings promise is really a monotonously repetitive pattern when spread across a wall of any size. Finally, there is something disturbingly dishonest about using a bogus siding that does not belong to the architecture of the house. No matter what modern siding may save you in future maintenance, it can never compensate for the destruction of a building's character.

REMOVAL

Unhappily, mass-produced siding on your house is miserable to remove. There is no telling what lies below, without serious excavation. If you are lucky, the original siding is still intact. More likely, however, the ornamentation has been sheared off, sections of the original siding have deteriorated, and caps have been filled in with scrap lumber. Therefore, do not remove modern siding until you have well-thought-out plans and sufficient funds to confront whatever the removal reveals. Sample exploration at enough different locations can give an experienced eye some clues as to the condition of the underraiment. Check above windows and doors, near ground level, near the roof line, and midwall. By all means make sure that the siding is not pulled off during the rainy season.

Removal of this siding is basically an unnailing operation. Vinyl and aluminum siding are the largest and thus the easiest to remove. Tarpaper brick and compressed stone cover less square footage per unit and so take more time to remove. Asbestos shingles must be removed one by one. Use a crowbar, or chisel apart the shingle enough to get a purchase on the nail with a claw hammer. Asphalt shingles are secured at the lower edge through one thickness only, so work from the top down. A dumpster within tossing distance and decent aim will simplify the cleanup operation considerably. Textured spray paint has to be sandblasted off.

The choice between repairing or replacing the underlying siding or shingles demands careful comparison of labor and material costs for puttying nail holes, caulking seams, and repainting the original, and the cost of removing the old siding altogether and starting again. The latter solution is often more economical, although neither remedy is cheap.

(CHAPTER 6)

WINDOWS

More mistakes are made in the restoration of windows than with any other part of the house. Windows are so very important to the appearance of the house that poor decisions can really detract from any other correctly restored area. There are three window designs appropriate to most Victorians: double hung, fixed, and casement. The frames for these windows should be wood; aluminum totally negates the true Victorian character.

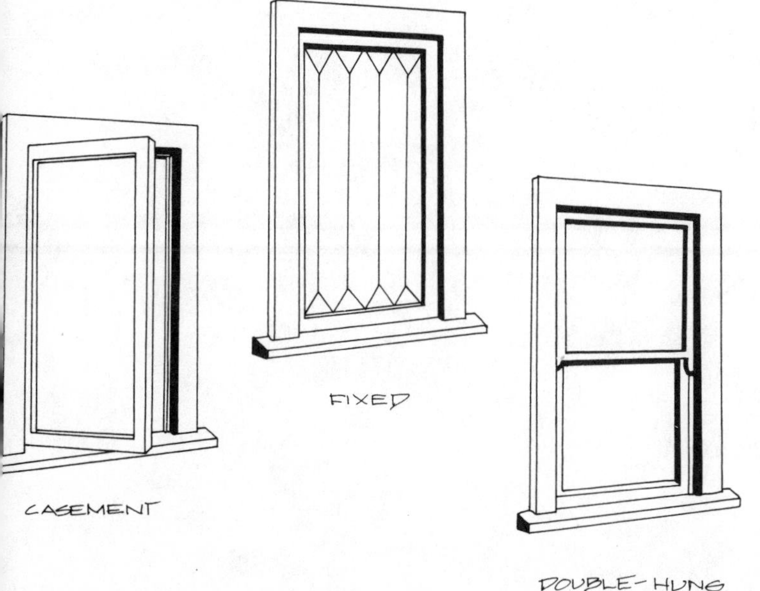

Types of wood frame windows

(DOUBLE HUNG WINDOW)

The double hung window opens with an up and down movement. An upper outside sash slides down, and a lower inside sash slides up. This is a practical design in the sense that it facilitates circulation, even when there is only one window in the room. The top and bottom sections can be open simultaneously, so that incoming air can enter the room, assume the inside temperature (usually warmer than outside), and then flow out through the open upper portion. The window movement is controlled by cord or chains on pulleys with weights, or by modern spring mechanisms concealed inside the jamb. Authentic double-hung windows are always made of wood and belong on a variety of Victorians.

With age, double-hung windows are subject to poor fit, broken sash cords, and fouled pulleys. These extremely bothersome defects are so surprisingly simple to repair that there is no justifiable reason for replacing a wood window with an aluminum one. To repair a double-hung window, first remove the trim with a hammer and chisel, keeping it intact for reinstallation. Dig into the void in which the movement mechanisms lie. If the cord is broken, lift the weight off the pulleys (watch your fingers) and install a new rope. If the weight is broken, that is, if it is split in two and half of it is sitting at the bottom of the void, with a screwdriver attach some other heavy object of equal size weight in proportion to the rope. If the window is simply sticking and all the mechanisms seem to be in order, check to see if the weight is distributed

Parts of double-hung window

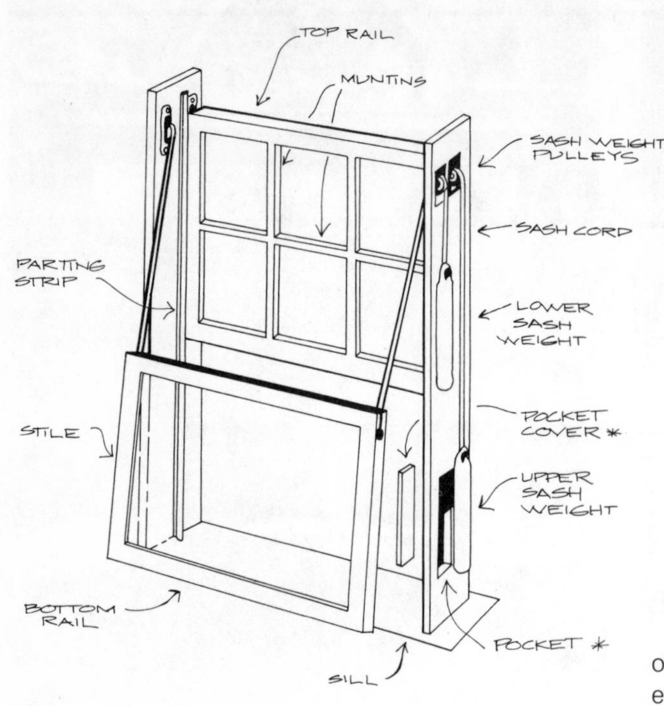

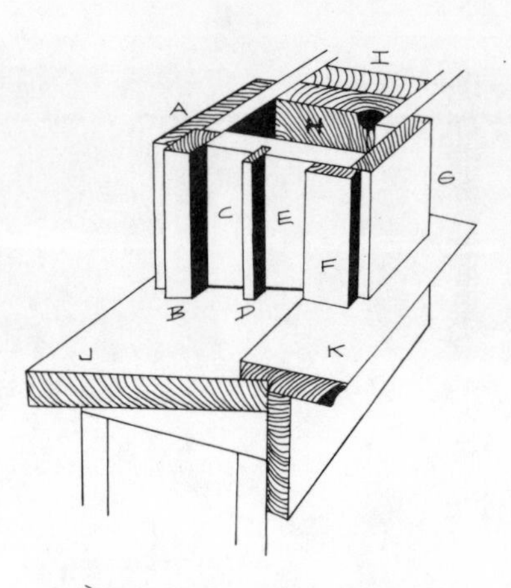

evenly; the cords should be of equal length, with the weights hanging uniformly. Another cause for a sticking window may be irregular house settlement.

[CASEMENT WINDOW]

Wood frame casement windows became popular about 1905. The casement window opens with an outward movement. It functions much like a door; it is attached to the frame by hinges along its vertical edge and swings outward to open. These windows seldom need repair, except for occasional tightening of loose screws that hold the handles and balky sliding rods. If you have to replace the handles, be sure and use properly sized screws; too-long screws may be driven into the glass and crack it. The aluminum frame casements are often very drafty because they do not fit tightly.

[FIXED WINDOW]

The fixed window is a pane of plain, stained, or leaded glass enclosed in a rectangular or specially shaped wood frame that does not open. Like a movable window, a fixed window's dimensions are carefully proportioned to the mass of the house and the other features of the facade. Fixed windows are found on all Victorian styles. (The leaded bars in leaded and stained glass may become bowed and require straightening.)

Double-hung windows: do's and dont's

ORIGINAL DOUBLE-HUNG WINDOW (ITALIANATE)
DO LEAVE IT INTACT OR REPAIR OR REPLACE IN-KIND. WINDOWS ARE A KEY INTEGRAL PART OF THE ARCHITECTURE.

OPENING BLOCKED DOWN TO ACCEPT STOCK ALUMINUM FRAME...
DON'T DO THIS. IT LOOKS MAKESHIFT AND MARS THE PROPORTIONS & APPEARANCE OF THE HOUSE.

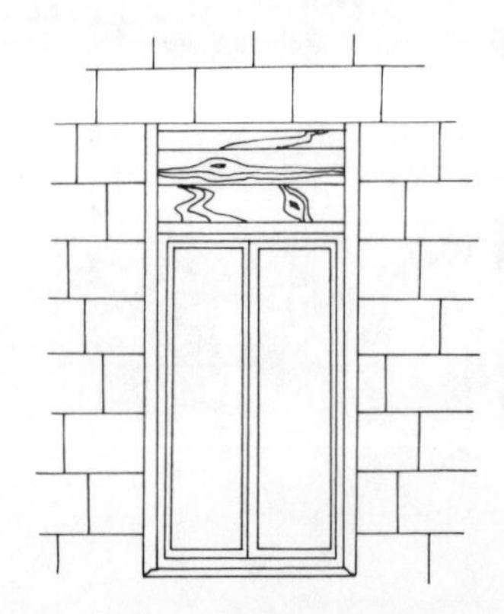

ORNAMENTATION REMOVED TO PUT ON ASBESTOS SHINGLES. NEVER DO THIS. TOTAL LOSS OF VISUAL INTEREST RESULTS.

WINDOWS

(REPAIR AND REPLACEMENT)

Two problems that plague all types of windows are broken glass and rotted wood frames. (The technique for replacing window glass is illustrated in Chapter 7, The Front Door.)

Wood frames are subject to deterioration from years of use, water accumulation, and insects, especially termites. Evidence of decay is spongy-feeling wood, brown or black discoloration near joints; or rippled, split, or flaking paint. Check the sash and sill for water stains resulting from condensation running down the glass.

Most people decide to replace an entire window frame because of a rotted sill. You can save money by just repairing the sill. The main idea is to fill in the holes and provide drainage toward the outside.

To build up a slope that will drain, use a penetrating epoxy marine product, carefully following the directions on the label. (More details are in Chapter 11.) For holes and cracks, first scrape away all loose material; then soak the sill for 1 day with a penta product, to kill the rot-causing organisms. Now saturate the sill with linseed oil. Wait another day, and fill the cracks and holes with putty. A few days later, prime and paint the wood.

For bad deterioration, use plastic wood or its homemade counterpart, a paste of sawdust and waterproof glue. Apply one or more coats, not more than ¼ inch at a time, and let each coat dry thoroughly between applications. Make sure the new wood slopes toward the outside. Sand, prime, and paint the wood. For irreparable deterioration, replace the sill entirely. Remove the old sill, being careful not to cut or chisel into the *stool,* which is the ledge on the interior side of the window frame. Use the old sill as a pattern for the new one. Sand the new sill, and bevel the edges slightly to ease installation. Anchor the sill to the window casing with brads, sinking the heads slightly below the surface of the wood. Putty the holes and seal them with shellac. Caulk the joint between the sill and frame. Finally, prime the wood and apply two coats of outdoor paint. (Refer also to the illustration How to Replace Door Threshold in Chapter 7.)

WOOD REPLICAS

If you have to replace your wood windows, compare the products offered by several window manufacturers. You can often obtain exceptional duplicates for a comparably lower price than those fabricated in small volume in wood crafting shops. A good window, aesthetically as well as functionally, is the double-hung facsimile made of pine or fir, coated with a white vinyl finish, and preglazed with energy-efficient glass. The preassembled window is an ideal compromise to the real McCoy because it saves much time. Not only is the painstaking synthesis of parts taken off your hands, but if you had planned to paint the window casing white, the white vinyl finish solves the problem. The vinyl finish requires no further painting maintenance and prevents water seepage better than paint. Some manufacturers also produce the vinyl-finished windows in other colors, but these hues are often raw and plastic looking, not gracefully blending with the Victorian design.

[ALUMINUM WINDOWS]

Although they are less expensive than wood windows, aluminum windows ruin the visual appearance of a Victorian. On any type of aluminum window, the pane of glass looks black when seen from a short distance, say, across the street. On a wood window, the frame and trim are flat and broad, providing a visual transition between the glass and the siding. The aluminum window eliminates the window frame, substituting a shiny filament of extruded metal. The result looks like a grotesque eye without eyelashes.

The shape of a prefabricated aluminum window is not the shape of a window opening on older houses. The Victorian Italianate, for example, was designed for a tall, narrow window. Because the standard aluminum product is not tall enough, a homeowner who uses one is forced to block down the house's window opening. Many people do this in an ugly manner, using plywood scraps and artificial siding. Similarly, the upper sash of an Italianate window is often curved or indented, and the rectangular aluminum frame will just not adapt to that shape.

Another visual problem with aluminum windows is the pattern of the panes. The division of window glass into panes is a conscientious design decision integral to the architectural style of the Victorian. The double-hung window is a one-over-one style; the Colonial Revival features the six-over-one (six panes to a window) and the eight-over-one (eight panes to a window) styles. Aluminum frame windows do not offer this option; in fact, they are designed just the opposite: The window is divided vertically by a reflective metal strip, instead of being divided horizontally by a substantial wooden strip.

A functional problem of aluminum windows is that they are not designed for optimum ventilation. Because they have only a single, sidelong opening, the air does not circulate unless there is another open window across the room and a decent breeze, and even so, this can be quite drafty.

Aluminum windows are obviously out of context with Victorian design. People use them to save money. Aluminum windows are cheaper, but is the saving worth the value lost in the house's integrity and resale attraction? Instead of considering the price of replacing an individual window, evaluate the cost in the context of the total restoration. For example, say you have 10 medium-sized, double-hung windows, of which eight are too rotted to repair and two have broken sash cords. You are tempted to replace all 10 with aluminum frames, saving 60 percent of the cost of comparably-sized wood windows (a difference of about $70 per medium-sized window). Replacing all 10 windows with aluminum instead of wood would save you $700. But if your total restoration budget is $10,000, the $700 is only 7 percent of the whole package, a fairly marginal economical difference considering the visual disaster that will occur. Instead, you should repair the sash cords on the two broken windows yourself and buy wood replicas for the other eight.

However, if you still feel that wood framing will overextend your budget, there are some reasonable compromises you can make regarding the use of aluminum windows. To reduce the negative aspects to some degree, consider the following ideas:

1. Keep the wood windows on the front of the house, but replace the side or backyard windows with aluminum. This keeps the street scene historically intact.
2. If you live in a coastal area, use anodized aluminum with an integral color like bronze or a baked-on color like white. This minimizes reflectivity and cuts down the probability of the aluminum surface pitting from salts. (Anodized aluminum windows cost a third more per unit than clear ones.)
3. Leave the wood trim in place, to maintain better visual balance.
4. Use the single-hung aluminum windows because they resemble double-hung wood windows. *Never* use the sliding type.

Stymie the temptation of using aluminum windows by properly maintaining the wood frame windows you already have. Keep the joints caulked, the holes puttied, and the paint in good condition.

[ENERGY]

Much heat loss and heat gain occur at the windows. Windows with a southern exposure warm a space; those facing north rob a space of warmth. Three parts of the window contribute to heat transmission:

1. *Frame and sash.* Wood is a better insulator than metal, so wood windows do not lose heat as rapidly as aluminum ones do.
2. *Glass.* Conventional glass transmits heat rapidly. Double-glazed windows minimize heat transfer to about half but cost more than twice as much as regular windows. If you are on a tight budget, consider double-glazed glass for very large windows with exposures to extreme weather conditions, such as a picture window that faces west and makes the room unbearably hot in the summer.

Double-hung Windows: weather strip replacement

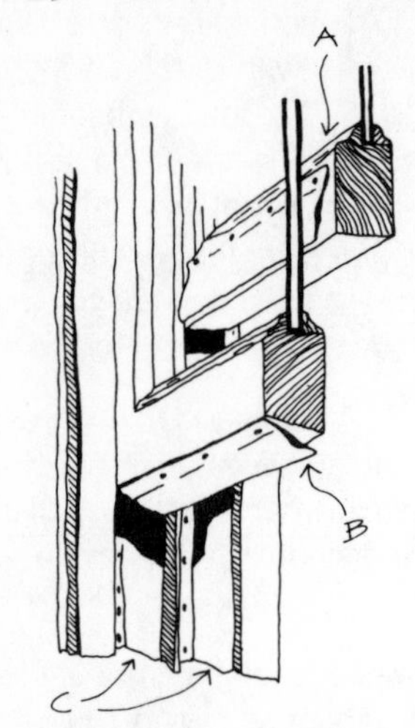

THIS SKETCH ILLUSTRATES THE PLACEMENT OF SPRING-METAL WEATHERSTRIPPING. FOAM RUBBER AND VINYL TYPES ARE ALSO AVAILABLE.

A) INSIDE OF UPPER SASH BOTTOM RAIL (SEALS GAP BETWEEN TOP & BOTTOM SASH WHEN WINDOW IS CLOSED)
B) BOTTOM OF LOWER SASH
C) SASH CHANNELS (DO NOT COVER PULLEYS)

Window security: vulnerability of aluminum frames

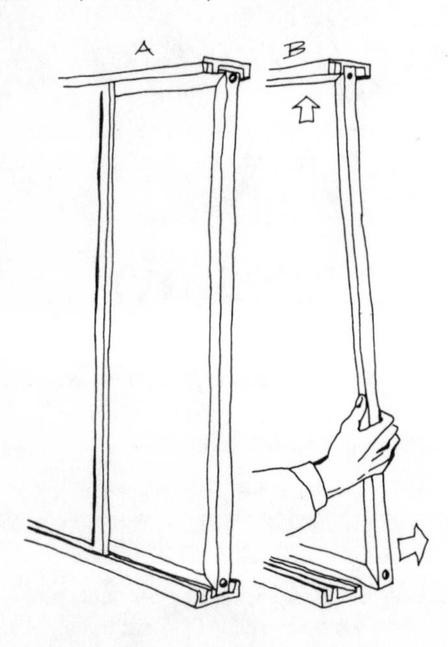

A
B
MANY ALUMINUM WINDOWS ARE SUBJECT TO EASY REMOVAL BY INTRUDERS. THE INSIDE (SLIDING) SASH FITS LOOSELY IN ITS TOP CHANNEL (A) & CAN BE REMOVED BY LIFTING UP WHILE PUSHING IN (B).

3. *Leakage.* Air leaks between the sash and frame, between the frame and the wall surface, and where the sash and rails meet. Weatherstripping provides a tight seal, eliminating drafts, and it is simple to install: just tack a piece of weatherstripping over any drafty seam.

(SECURITY)

Aluminum windows are much less secure than wood frame windows. Aluminum windows fit so loosely in their frames that they can be removed completely by lifting the window up and pushing in. (This is a deliberately designed feature, to enable cleaning from both sides at high elevations.) And locking aluminum windows is difficult. Double-hung wood frame windows have a turn-locking mechanism that prevents separation of the upper and lower components. Wood windows can also be equipped with key-operated locks for greater security. Windows are extremely vulnerable to breakins, and it is difficult to completely burglarproof them because an intruder can just break the glass. If you plan to add security bars to any windows, equip them with interior emergency-release mechanisms (often required by law).

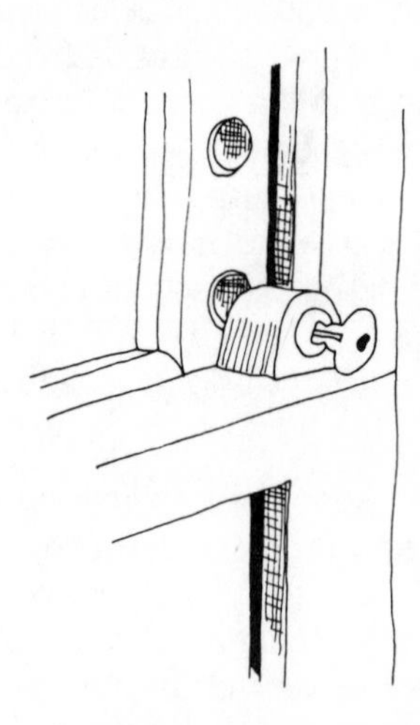

Window security: keylock for double-hung windows

(CHAPTER 7)

THE FRONT DOOR

Visually, the front door catches the eye, and functionally, it protects against unwanted intruders. The importance of the front door was recognized on older homes by the care given to the design. The front door was always a panel door, constructed of top-quality hardwood and detailed with moldings, proportions, and architectural features appropriate to the rest of the house.

Expensive Victorian homes had a pair of centrally opening golden oak doors, with a panel of beveled or flashed glass on top and recessed molding of veneered, grained mahogany below. More common Victorians and Colonial Revival houses had a single broad oak door, with a large window on the upper portion and decorative molding below. The front door of many Victorians was made of fir, redwood, or oak, consistent with the wood selected for the interior. If the door had a window, it was leaded, made of small panes, or beveled.

Country Victorians had an oak door stained to match the finish of the interior hardwood. The fancier Period Revival home had front doors stained and textured to look rustic.

Unfortunately, these valuable doors are often disposed of in the course of home improvement. Substituting a modern flush door for a beautifully paneled original is a pathetic waste of an architectural resource as well as money, because most front door problems can be corrected. There are four main problems with old front doors: (1) fit, (2) wear, (3) hardware, and (4) security. (Fit is more pertinent to interior doors and so is discussed in Chapter 13.)

(WEAR)

The front door, like any surface common to the exterior, gets a great deal of abuse from the sun, wind, rain, and snow. It also has problems because it is a movable object, subject to wear and tear from the chronic impact of twisting and turning knobs and from being slammed, kicked, and scratched by people and animals. Such use puts more stress upon the front door than on many other elements of the house.

PANELS

Many homes originally had paneled front doors. The shape and number of panels per door varied, but the construction technique was the same: slender panels enclosed in a heavy structural framework and secured in place by moldings, all imparting a solid, three-dimensional design. The lower panels were made of wood, the upper panels of etched, leaded, or stained glass. The glass was rectangular, round, or oval; it was in a large, single window, in a matching pair of windows, or in a composition of square panes.

VENEER

The framework of the door, the portion that surrounds the panel, is typically a solid piece of softwood with a thin veneer of oak, walnut, or mahogany on either side. When moisture seeps between the softwood and the veneer, the two layers separate, and the veneer begins to fray at its lower edge.

1990

How to replace door glass

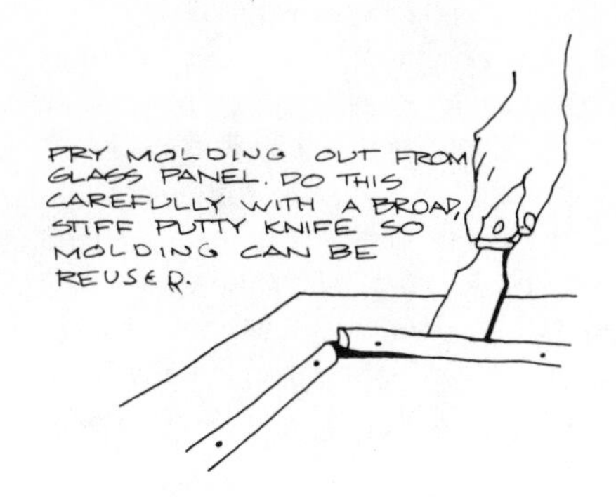

PRY MOLDING OUT FROM
GLASS PANEL. DO THIS
CAREFULLY WITH A BROAD,
STIFF PUTTY KNIFE SO
MOLDING CAN BE
REUSED.

PLACE DOUBLE-STRENGTH
PANE CUT SAME SIZE AS
ORIGINAL (OR 1/8" SMALLER
THAN OPENING IN HEIGHT &
WIDTH), IN BED, PRESSING
TO FORCE COMPOUND UP
AROUND EDGES. APPLY
2ND LAYER OF COMPOUND
AROUND GLASS EDGE.
RENAIL MOLDING
FILL IN HOLES.

To repair this condition without making it look like a patch job, repair all the way across the bottom of the door for a horizontal condition, or all the way up to the top of it for a vertical problem. Cut off the ragged piece of veneer to get to the sound softwood; use a backsaw or an electric router with a straight blade set at a shallow depth. Glue a new piece of veneer of the right size, color, and texture into place, and clamp down until the glue sets. Consult your local lumber yard or hardware store for the proper glue or mastic to use; it is as critical as the veneer itself.

If only a small corner of the veneer is damaged, chip it back with a chisel. Fill it in with a wood putty of matching tone. Repairs are even simpler if the door is already painted, because touch-up paint or a new coat can mask corrections that are hard to conceal on a natural wood finish.

To ward off future damage from scuffing, install a kick plate (use a drill and screws). Try salvage yards for interesting brass plates that match the other original features of the door. Otherwise, use a very plain piece of nonferrous metal or a heavy-duty plastic piece of a neutral or matching color. Do not use brightly colored plastic, aluminum, or stainless steel because these materials are too shiny, scratch easily, and are out of context with the Victorian's character. Copper and brass are good choices because with age they acquire a patina, or they can be shined to a warm glow. The plate should be 2 inches narrower than the door width, to provide for doorstops.

PAINT

A natural wood door is preferable to a painted one because its textured grain lends richness to the appearance of the house. If the original door is in decent condition and is unpainted, perpetuate the natural finish. However, if the finish is severely weathered or repairs have left ghastly scars, painting is a much better solution than installing a flush door and cheaper than buying a new one that matches the style of the original door. If you paint the front door, be sure to first treat the surface with a wood sealer or the paint will be sucked up, and even three or four coats will not achieve a finished look. The seal also provides protection from future moisture penetration. (See Chapter 9 for painting instructions.)

If the door has already been painted and you want to reverse the process, see if you can strip the door yourself, or hire a professional who does not use the vat process. Dunking wood into a tub of solvent indeed removes the paint, but the strong chemicals also saturate the wood and destroy much of its strength and natural luster. Use a brush-on and scrape-off technique to minimize excessive damage to the wood. After the paint has been removed, treat it with a varnish or natural wood oil. (Refer to Chapter 11, Interior Walls and Woodwork, for more information about stripping.)

ORNAMENTATION

Ornamental detail is extremely important because the subtleties found in it often distinguish one similar style from the next. Original Victorian doors had decorative wood pieces applied over the lower panels in a design that matched the ornamentation used elsewhere on the building. By no means eradicate these decorative elements; patch and remold them if at all possible. If the wood is irreparably decayed, visit a local historical supplier and try to buy molding that most clearly matches the original. Never substitute plain molding or ornamentation that conflicts with other details of the house.

THRESHOLD

A threshold creates a tight finishing seal between the interior and exterior of a house. Although made of hardwood (usually oak), the threshold receives extreme wear and abuse from being stepped on, scuffed, and tripped over. A worn threshold gives the entryway an unnecessarily shabby appearance. To replace the threshold, first swing the door wide open. If the door does not clear the threshold, pull out the hinge pins and remove the door. Using a prybar, start pulling the doorstop at a place ½ inch away from the door jamb; start at the bottom and proceed to the top (see illustration). Then remove the stop completely. Do not cause any unnecessary holdups by splitting the stop.

Now try to pry up the threshold intact, to use as a pattern. If necessary, split the threshold into two pieces by using a wood chisel. Place the threshold on a board of

How to replace door threshold

SWING DOOR WIDE OPEN. IF IT DOESN'T CLEAR THRESHOLD, PULL HINGE PINS & REMOVE DOOR. USING A PRYBAR, START PULLING DOOR-STOP 1/4" AWAY FROM DOOR JAMB, FROM BOTTOM TO TOP, THEN RE-MOVE STOP COMPLETELY.

TRY TO PRY UP THRESHOLD INTACT FOR USE AS A PATTERN. IF NECES-SARY, SPLIT INTO PIECES WITH A WOOD CHISEL. TRACE OUTLINE OF THRESHOLD ON BOARD OF EQUAL THICKNESS (OR TAKE CAREFUL MEASURE-MENTS FIRST IF YOU MUST SPLIT THE OLD ONE). TRIM NEW THRESHOLD TO SIZE, THEN TAP LIGHTLY INTO PLACE.

FOR HARDWOOD THRESHOLDS (THE BEST CHOICE FOR WEAR), DRILL PILOT HOLES A BIT SMALLER IN DIAMETER THAN NAILS TO AVOID SPLITTING WOOD. COUNTER-SINK NAILS (OR SCREWS). FILL IN HOLES WITH PUTTY.

equal thickness and trace its outline, and then cut an identical piece. However, many thresholds are not of a consistent thickness; they are wedge shaped, milled and angled much like a windowsill. If you have to replace this type, buy a prefabricated one at a hardware store or building-supply shop.

Before setting the threshold back into place, be sure that the groove between the finished floor and the exterior has been scraped free of debris and scum. Then seal the seams, which might otherwise let moisture seep through. Also caulk the underside of the front tapered portion of the threshold before tapping it back into place.

When securing the threshold, first drill pilot holes slightly smaller in diameter than the nails, to avoid splitting the wood when you are nailing. Countersink the nails, and fill in the holes with wood patch or putty. Finish the threshold with varnish or oil; it will last 50 or 100 years.

Do not substitute an aluminum threshold. It is much cheaper than an oak threshold, but its appearance will cheapen even the most beautiful front door.

[HARDWARE]

The doorknobs and hinges on an original Victorian front door are usually made of brass, steel, or brass-plated steel. Whether elaborately detailed or plain, the brass takes on an elegant luster when maintained. Most older front doors are hung on heavy-duty, butt-type hinges made of two leaves, screwed to the door and the door frame and held together by a strong vertical pin. Often the weight of the door enlarges the holes that once held the screws snugly. This can result in a stubborn door that sticks on one side. To tighten the hinge attachment, remove the door by knocking out the hinge pin with a nail and hammer while the door is closed. Then remove the hinges. Next, put the hinges back on by using longer screws; or refill the screw holes with putty, plastic wood, a dowel, or wooden matchsticks, and reset the screws.

If the hinges are covered with paint, either soak them in a coffee can of paint remover to return the metal to its distinctive natural finish, or scour the oxidation with a brass tarnish remover, following the instructions on the label. Coat the hinge with a clear acrylic lacquer, and thereafter, to highlight the detail, polish the brass every few months for about a year. Apply the same treatment to the brass mail slot, if there is one.

Original doorknobs are beautiful to behold and feel luxuri-

ORIGINAL DOOR KNOB AS HANDLE WITH AUXILIARY LOCK

Door security

2070

ous. You should retain them for use as handles, even if their lock mechanisms are inadequate for security. Keep the faceplate tarnish free with a brass cleaner. If the brass plating has worn off in spots, replate the faceplate. The brass cleaner will also keep rust off the exposed steel.

If the latch bolt is unresponsive, fix it for the convenience of having a door that closes properly (rather than for security). Sometimes the latch bolt and the steel strike plate do not meet because the door itself does not fit properly. To determine if fit is the problem, see if the hinge screws are loose, as described; if they are not, conduct the experiment outlined in Chapter 13, on doors. If the problem is the latch mechanism, detach the latch from the door and have a locksmith repair it.

[SECURITY]

Today security is more important than ever, but sometimes the solutions are at odds with a Victorian's architectural style. The following recommendations demonstrate that you can make your house secure without sacrificing its appearance.

Remember, home security extends beyond the front door. It includes all other openings into the house, such as other doors and windows, and the visibility of the house from the street, the presence of street and yard lights, blind spots behind walls, and even the number of neighbors casually looking out their windows. In fact, more breakins occur by entry through windows and the back door than through the front door. The back door is usually of flimsy construction, and because it is "hidden," intruders prefer to use it.

For the most efficient security system, direct attention to the back and side doors, garage door, windows, and exterior lighting. Also consider installing an electronic burglar alarm system, especially if you live in a high-crime neighborhood. A system may be a metal box, fire bell, or lights by the front door; such conspicuous elements often deter a thief. Or you may want to have one or two watch dogs. Locks and mechanical systems are not as foolproof as the keen hearing sense and the barking of a dog.

DEADBOLTS AND BARS

When providing security for the front door, the real consideration should be the vulnerability of the door itself. Modern, flush, solid-core doors are harder to break down than the original paneled ones, but the modern ones are so inferior aesthetically, and there are so many aspects to security, that retention of the original panel door is highly recommended.

There are several security improvements that are very effective without obliterating the original design. Old doors have mortise locks that are substandard for today's security requirements because they do not have enough throw. The minimum acceptable throw is 1 inch. Keep the original metal knob and escutcheon plate, using them as ornamental handles. Supplement them with a twin-cylinder, dead-bolt lock with a 1-inch throw. This type of lock requires a key both inside and out, so even if someone breaks the door's glass and reaches in, he cannot open the door. Be sure to keep a spare key nearby to use for emergency exits.

Reinforce the wood panels by installing plain metal bars on the outside. Secure the bars with throw bolts so they cannot be unfastened from the outside. Paint the bars to blend with the color of the door. Another technique is to add a piece of plywood to the door. Fasten false panels over the plywood (using countersunk screws or nails), and paint them to match the door.

Some Victorians have paneled double doors that open at the center. The weak security is at the point where the doors meet. The best solution is to install a vertical rod lock, which attaches the door to the floor and ceiling when fastened into place.

DOOR WINDOWS

Front doors that have glass panels invite people to peek into the house. Install a translucent curtain or shade, or sandblast or frost the glass with a spray. Another idea is to tack a wire screen onto the interior or exterior of the window. Place molding around the seams (as suggested for plastic and metal grid covers, which we discuss in the following paragraphs). All these alternatives provide adequate light yet control visibility from the exterior.

Italianate double doors

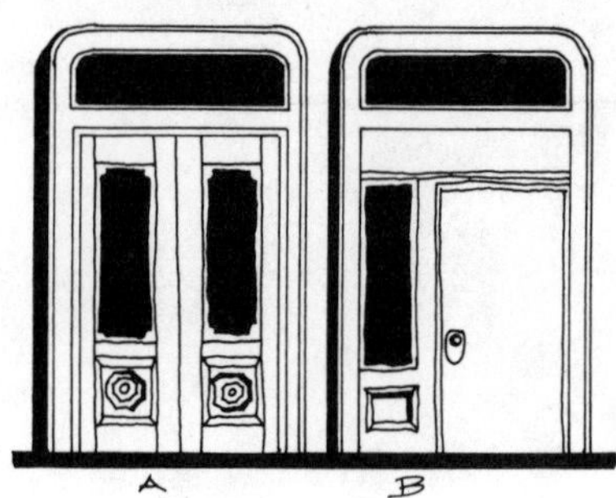

A) ITALIANATE DOUBLE DOORS LIKE THE HOUSE STYLE, THEY ARE VERTICAL IN EMPHASIS & RICHLY 3-DIMENSIONAL. ORIGINAL DOORS WORK WITH YOUR HOUSE'S ARCHITECTURE. IF YOU HAVE THEM, KEEP THEM. IF THEY NEED WORK, FIX THEM; IF THEY HAVE BEEN REPLACED, IT'S WORTH THE EFFORT TO FIND NEW OR SALVAGED REPLACEMENTS SUITABLE FOR YOUR HOUSE STYLE.

B) BLOCKED-DOWN REPLACEMENT SUFFERS FROM ASYMMETRY, SQUATTY PROPORTIONS, & LACK OF VISUAL TEXTURE. AVOID THIS ALL-TOO-COMMON REHAB ERROR.

For security, reinforce the glass by adding a panel of break-resistant plastic (like Lexan MR 4000) in front. A 2 x 3-foot piece, 3⁄16 inch thick, is virtually bulletproof, as are certain types of plexiglass. Real bulletproof glass is expensive, extremely heavy, thick, and imposes upon the door's appearance. To fasten the plastic to the door, first drill 1-inch holes through the plastic and into the door. Be sure you precisely align the holes in the plastic with the door holes, because plastic is not as easy to patch as a door. Then fasten 1-inch screws tightly and securely.

The resulting appearance is a bit obnoxious because the shiny screws look like rivets or sequins. Disguise the seam by installing a flat wood molding over the rim. Drill correctly, aligning holes through the wood and the plastic (but stagger the holes so they do not interfere with the screws). Countersink finishing nails as fasteners, and fill in the gaps with putty. Then stain or paint the molding to match the door. If the edge of the plastic peeking through this wood sandwich bothers you, tack on a wood finishing strip that abuts the plastic and wood edge. This should conceal all the fastening techniques and establish a secure panel sympathetic with the design.

Another alternative window security procedure is to shield the window in front with a woven metal screen of #10 wire on a 1½-inch grid. Conceal the screen edges with the wood molding as just described, but be sure to first paint or treat the molding so it does not rust (a clean finish or Rustoleum works well).

DOOR FRAME

No matter how secure the door and lock, if the frame of the door is vulnerable, forced entry is fairly easy. Many door frames are set into the wall, without solid connection to the studs. Since the lock slides into the frame, which is just a ¾-inch piece of wood, the frame is all that holds the door shut. A solid kick to the frame at the lock will rip the bolt out of the frame or the frame out of the wall. Remedy this situation by adding a solid wood blocking member equal to the height of the door between the door frame and the stud. Nail them solidly together to create a column of strength. This will also lessen the tendency for the plaster or gypsum wall finish adjacent to the frame to crack, buckle or cave in.

The door frame is also vulnerable to lateral pressure, the effects of which are evidenced by the outward bulging and bowing of the frame and its supporting wall studs. The straightening-out operation is not difficult at all, but a little messy. Remove the door molding from the frame (saving the molding). Pull off the plaster or gypsum wallboard past the stud to the second or third stud away from the frame. Also remove any lath or interior sheathing. Remove the door frame (it resembles a pair of cowboy's legs), and replace it with dry, vertical stud work. Then add horizontal 2 x 4-inch blocking between the other exposed studs at the lock level. Toenail the blocking in place, and resurface the wall (see Chapter 11). This bracing should very well take care of the problem and ward off any future ones.

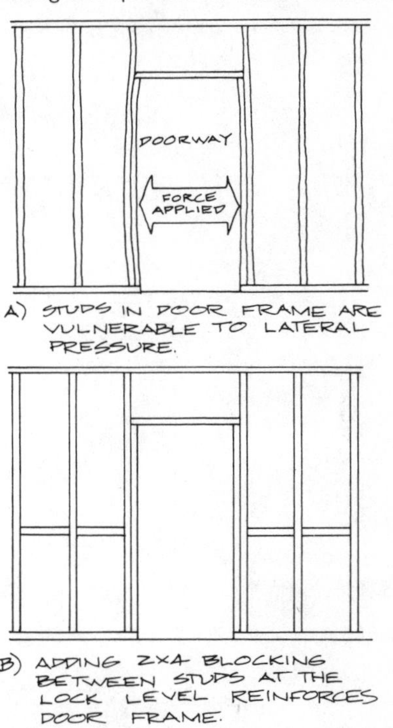

DOORWAY
FORCE APPLIED
A) STUDS IN DOOR FRAME ARE VULNERABLE TO LATERAL PRESSURE.
B) ADDING 2X4 BLOCKING BETWEEN STUDS AT THE LOCK LEVEL REINFORCES DOOR FRAME.

Door security: braced studs

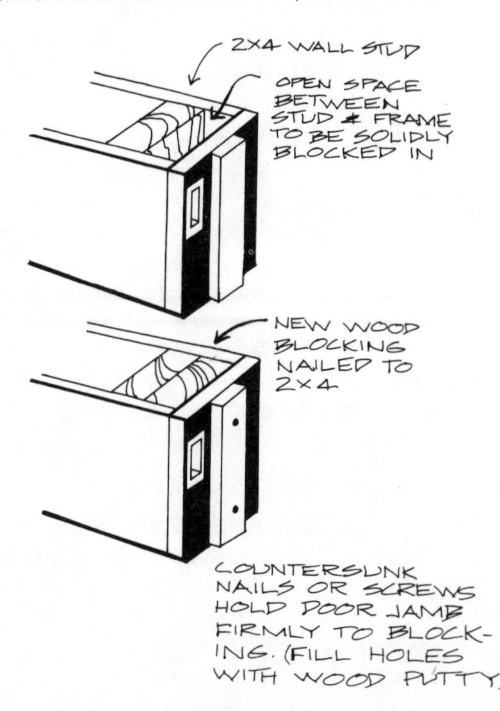

OPEN SPACE BETWEEN STUD & FRAME TO BE SOLIDLY BLOCKED IN

COUNTERSUNK NAILS OR SCREWS HOLD DOOR JAMB FIRMLY TO BLOCKING. (FILL HOLES WITH WOOD PUTTY.)

Door security: reinforced door frame

HINGES

Hinges are essential for maintaining a door's strong and secure attachment to the frame. They should be mounted on the inside of the door; if they have been mounted on the outside, they should have nonremovable pins. Replace modern hinges with the type that comes with a nonremovable pin. Do not sacrifice the original

Door security: Nonremoveable Hinge Pins

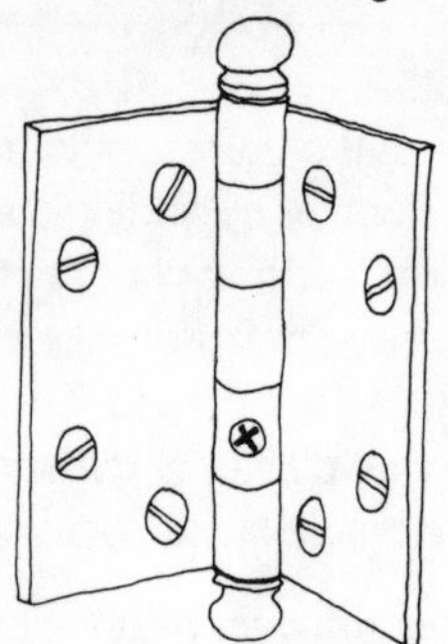

OPEN DOOR. FROM INSIDE OF HINGE, DRILL A PILOT HOLE INTO HINGE PIN. COUNTERSINK HOLE SO SCREW HEAD WON'T INTERFERE WITH CLOSING OF DOOR. INSERT SELF-TAPPING SCREW.

Door security

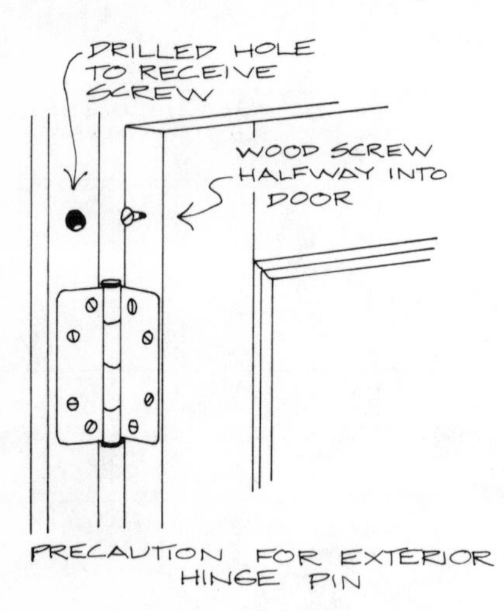

hinges in the name of security; you can make the pin nonremovable by drilling a hole at right angles through the hinge pin from the inside and inserting a machine screw. Another precaution is to install a wood screw halfway into the door near each hinge. Drill a hole into the jamb opposite each screw to receive the projecting screw heads. The door should stay in place even if the hinges are removed.

LIGHT FIXTURES

Front door security can be increased by attending to the environment around it. Exterior wall or porch ceiling-mounted light fixtures help spotlight the display of the entry, maintaining the warmth and strength of the entry when natural light is not present to articulate its richness and detail. And intruders are far less likely to go where they can be seen.

When choosing light fixtures, carefully discriminate. If there already are fixtures on the porch ceiling or house wall, do not assume that these are as original as some of the other elements, because many older homes were designed before electricity was a common household installation. If you notice that your fixtures are "dead" decorations, examine them closely; they may be equipped for gas lighting. If so, chances are that you have original lamps. To preserve these treasures, disconnect each lamp from its feedline, and remove the mechanisms. Keep the globe and as much of its metal base as possible. Have a reputable electrician reinstall an electrical system of socket, wires, and switch control, carefully adapting the new to the old.

If your present fixtures look decidedly "old," and if they have a design and detail sympathetic to the architecture, they may very well be original if your home was initially equipped with electricity. If you spot a "moon" globe or flat and flush contemporary lamps, do not save them. However, new light casings indicate fairly new wiring, which saves you spending time and money for an electrician's services. Research the type of fixture that was original to the architecture and replace it with a reasonable facsimile. Do not shy away from imitation antique lamps when it comes to porch lighting. This is an area where skimping on cost and being a little phony will not harm the architectural integrity of the house. It is foolish to spend a great deal of money on a valuable antique and leave it unguarded outside. Secondly, the lamp directs more attention to itself at night when lit. The major portion of the fixture that is accentuated is the globe. But whether the globe is made of glass or heat-resistant plastic, people rarely stare directly into the lamp, touch it, and examine its casing and metal base and other details.

[NEW DOORS]

If you cannot salvage the original door and so have to buy a new one, buy one that is secure and attractive rather than a flush, solid-core type that is incongruous with Victorian styles. Standard 3x6-foot x 8-inch flush doors made of particle board with a mahogany or birch veneer cost about $35 to $45, whereas for only $20 to $100 more you can get an authentic full-sized panel door at a salvage yard. And for about the same price you can buy a new paneled door that resembles an old style.

We highly recommend this solution because your restoration will then retain both style and structural stability. When you buy these exterior doors, insist that the panel portion be at least ½ inch thick, for security purposes.

Whatever type of door you choose, make sure it is the correct size for the door frame, because Victorian frames were often taller than those for modern doors. Otherwise you will have to assault dreadfully the appearance of the house by blocking in the leftover space. Search salvage yards and inquire at local Victorian restoration suppliers to obtain the exact proportions and style of door. Consider the importance of the door's relation to the facade of the house.

[CHAPTER 8]

STAIRS AND FRONT PORCHES

[STAIRS]

Exterior stairs are different from interior stairs as to the kinds of exposure and wear they receive. Outside stairs receive constant use and abuse from pedestrians and weather. More durable materials like concrete and flagstone hold up best, but they are often most inappropriate to the Victorian style. If the original stairs on your house have thoroughly deteriorated, replace them with the original material, in the original style. You can determine the accurate style by checking nearby Victorians or by referring to architectural historical books. Similarly, if the previous owner rebuilt the stairs in a non-Victorian style or material, one inappropriate to the house, you may want to replace them in the original style also.

WOOD

Deteriorated wood stairs are an extremely common problem with Victorian and Colonial Revival houses, but in order to preserve the building's architectural integrity, you should replace them with wood rather than concrete. After all, the wood steps on many a Victorian have lasted 90 years, so even if new wood steps last only another 90 years, they are perfectly adequate for your needs.

The greatest source of wooden stair problems is poor drainage; even the smallest accumulation of rain can begin the rotting process. The accumulation may be caused by one factor, or a combination of factors:

1. *Solid boards installed perfectly flat.* Water has trouble draining, so it will remain on and penetrate into the surface of the tread.

Stair dimensions for new construction

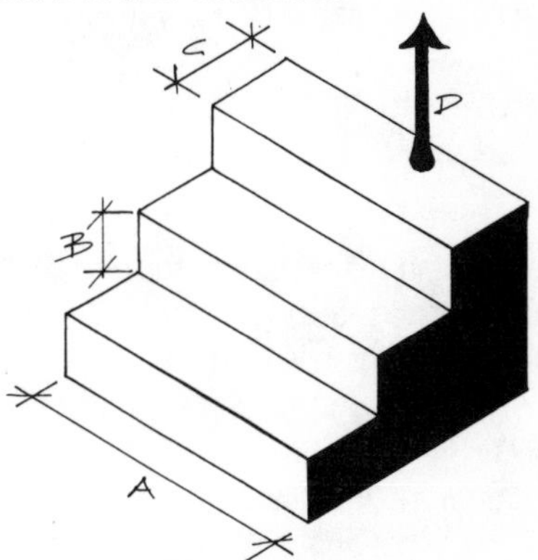

2. *Worn treads with a depression in the middle.* Water sits in the porous, worn center, creating an incubation puddle for decay. When repairing wooden steps with this problem, consider turning over the work treads. A shallow depression on the reverse side will have no detrimental effect on the step, and flipping the board will save you money and time.

3. *No paint and caulk.* Bare wood and unsealed joints permit complete moisture penetration.

4. *Drainage and drip paths on the stairway.* Other areas of

Improving drainage on wood steps

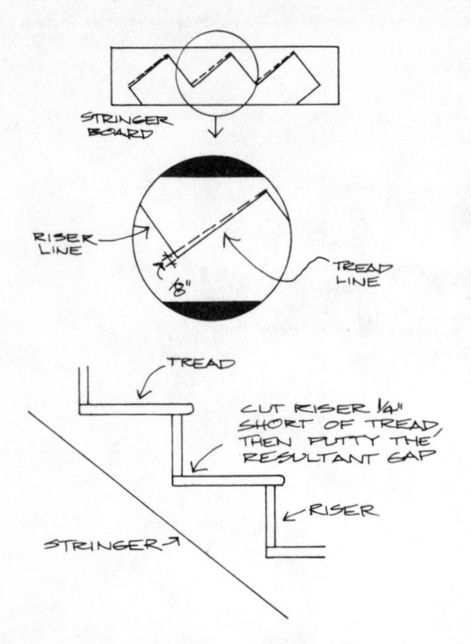

the house, such as roof gutters and downspouts, sometimes dump water onto the steps instead of diverting it properly.

Install new or recycled boards at a slight angle, say, ⅛ inch from front to back, as illustrated.

Prior to installation, coat the treads with a wood sealer; the end grains are critical areas. Make sure the tread nosing projects ¾ to ¼ inch over the riser, to keep the joint waterfree and to provide a shadow that is a definite design asset.

Do not direct drainage water onto the sill plate below the steps; direct it onto a masonry surface that slopes away from the sill. Design the railing so that the balusters are attached to a handrail on top, and to a freestanding shoerail on the bottom. This space allows water to run off the edge after a severe storm. (Refer to the handrail illustration in that section.)

BRICK

One problem with brick stairs, as with brick foundations, is settlement caused by shifting soil. To correct this fault, concrete must be injected beneath the existing structure, and additional bracing of steel reinforcement may be required. This is a difficult operation that sometimes calls for equipment more sophisticated than house jacks and should be executed by a qualified contractor.

Brick stairs, like brick foundations, are also subject to cracked mortar joints. These require immediate attention to prevent moisture penetration, which further cracks and rots the mortar. (To repoint mortar joints, refer to the section on brick foundations in Chapter 3.)

CONCRETE

Long concrete stairs are commonly found on Prairie School Houses and California Classic Bungalows. Shorter ones, often of only a step or two, are typical to the Period Revival Style. Occasionally concrete steps are embellished with redwood strips, mosaics, or tiles, and Mediterranean and Provincial houses are known for their shiny red-painted concrete steps and matching walks. Green and gray are popular paint colors for bungalow steps. You should stick with these colors because repetition lends visual unity to a neighborhood.

Grooves are another design aspect of concrete steps. They may be geometrically shaped in abstract patterns, or they may have parallel lines for skidproofing. You should keep the original arrangement of grooves if you repair the concrete, or reinstate the grooves if you completely replace the stairs.

Concrete steps are subject to *spalling,* which is the chipping off of small pieces from the nose of the tread. You can repair minor damage by using a high-strength mortar to bond new concrete to the old. If the damage is severe, consider adding a steel angle plate with an antislip surface. The steel plate has a textured rod (see the illustration) that must be firmly secured to the existing concrete. If it is not, the plate will dislodge and permit moisture to be absorbed into the joints.

Small cracks in concrete steps are normal and can be

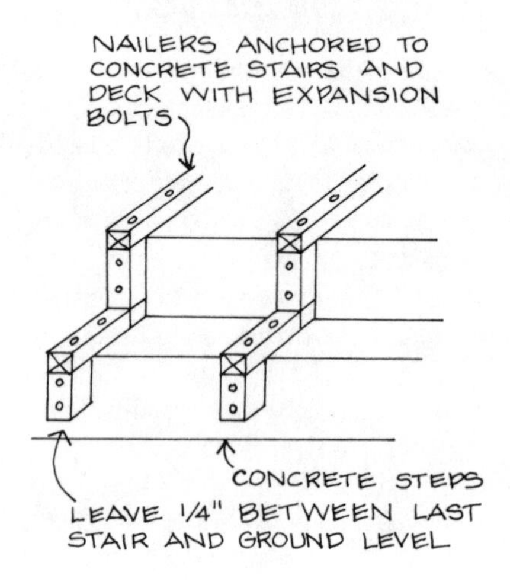

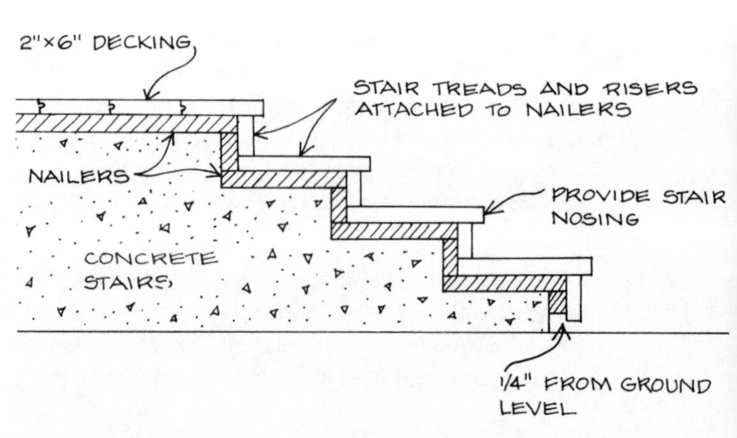

Repairing concrete steps

Steel nosing on concrete steps

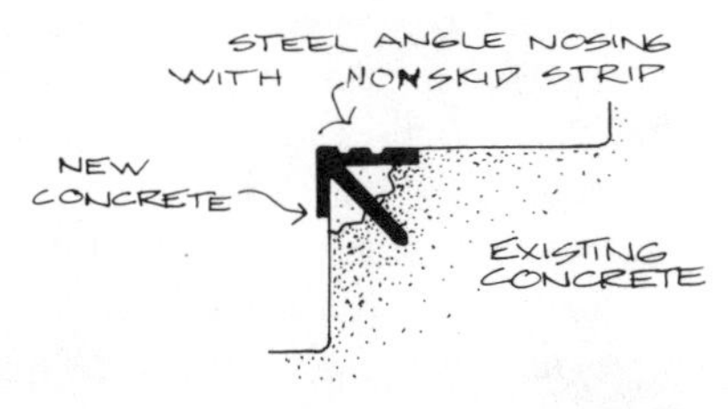

repaired with concrete patching cement. Major cracks, as in foundations, are signs of uneven settlement and should first be examined by an engineer.

HANDRAILS

In addition to being a decorative element, the handrail serves both structural and safety purposes. Many Victorians and shingle-style houses, however, were built with handrails that are inadequate for today's safety standards.

Most codes warrant handrail heights of from 30 to 34 inches above the surface of the finished tread; and stairways wider than 88 inches must have an intermediate rail for every 88 inches of width. Every handrail on a residential stairway must also extend at least 6 inches beyond the top and bottom risers and be finished with newel posts.

The sub-standard railing can be reconstructed to both

Stair railing construction

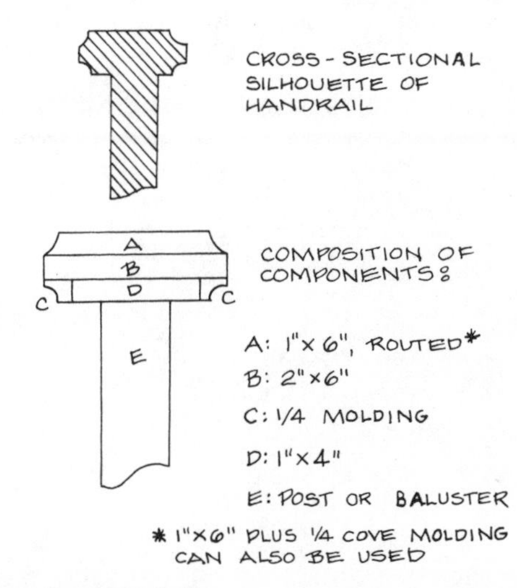

Constructing a handrail

code and architectural integrity. First, consider the original material from which the handrail was made, and by all means try to match it. Most Victorian homes have wooden rails that are solid and ornately sculptured. To reconstruct the handrail's silhouette, examine the proportions in cross section, and measure every dimension of the outline. This helps you see the individual components so that you will replace them as faithfully as possible. Most of these types of handrails are made from layers of standard-sized milled lumber, embellished with various types of moldings.

If the railing height is too low, sometimes boosting the railing will solve the problem. Do this by firmly anchoring a 4 x 4-inch minipost, cut at the railing's angle, to the inside of the stair stringer. Cut away the portion of the tread through which it imposes. (Be sure the fit is snug and properly sealed at the joints.) Extend the 4 x 4 up high enough to compensate for the required height minus the rail, which is minus 1½ inch. The 1½ inch is to accommodate a 2 x 2 shoerail placed on top of the 4 x 4s projecting out of the treads. This shoerail gives the entire element a balanced and finished appearance and permits excess water to flow easily over the edge of the tread.

The Colonial Revival handrail often poses a few problems when both reconstruction and bringing it up to code are involved. The handrail should be stepped, not diagonal. The stepped railing is a series of graduating blocks that in themselves compose small structures. Each step block

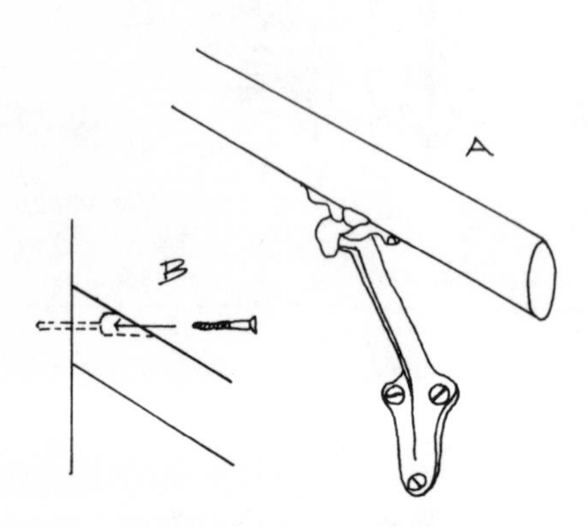

HANDRAIL DETAILS

A) BRACKET WITH CLOSET POLE RAIL (IF A 2X2 IS USED FOR THE HANDRAIL, CHOOSE A SMOOTH-FINISHED BOARD WITH ROUNDED EDGES.)

B) ATTACH THE UPPER END OF RAILING TO WALL OR PORCH WITH A COUNTERSUNK WOOD SCREW. DRILL PILOT HOLE FIRST.

Exterior handrail reconstruction

EXTRA HEIGHT CAN BE GAINED FOR CODE COMPLIANCE FROM 2x4 EXTENSIONS.

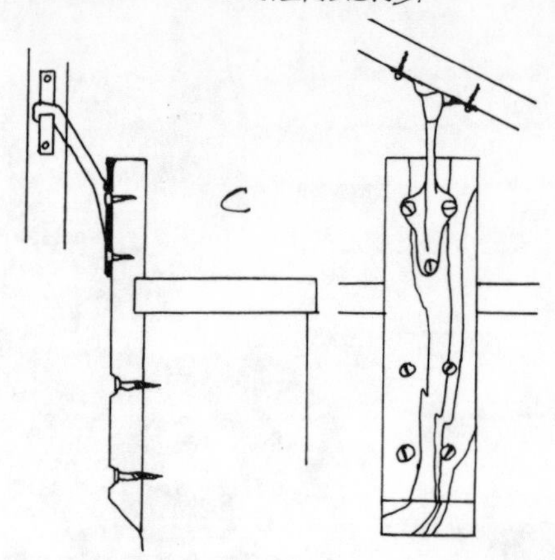

Adding height to substandard rail

consists of a framework, siding, and a cap. When trouble is present, it is not always so easily spotted as it is in an open rail. Rotten wood may be within the framework, which is covered by the siding and cap. A reasonable guess as to the condition of the concealed wood can be made by examining the siding. If the planking or cap is warped, checked, or decomposed in any way, chances are the framework is in no better shape.

To remedy the trouble, remove the surfacing and the other rotten boards and replace them with new ones of the same dimensions. The new boards and surfacing should be of redwood, treated with a preservative. If the previous siding was rotten from resting on bare earth, excavate slightly around the area so that about 6 to 8 inches of concrete are exposed. Clear away the earth so it will not settle back into the same position from drainage. Attach galvanized flashing to the wood around the lower area, but generously overlap the concrete, to direct most of the water outward instead of into the framework. Then replace the siding, ensuring that its bottom edges do not touch the ground. Caulk and silicon seal the inside seams where the siding meets the tread or landing.

Also replace warped caps. Be sure to maintain the same amount of overlap as with the original, and reinstate any finish molding.

The step rail design is heavy, imposing, and deliberate. Do not fudge or force it to meet rail-height codes. Instead, let its original integrity be; add on an interior oak rod bannister. Anchor it onto the inner face of the step rail (about 30 to 34 inches from the tread) with brass hardware. Finish it with an inconspicuous tone. This solution enables the staircase to uphold its architectural indentity as well as meet current safety codes.

Brick handrails always accompany masonry stairs. Fortunately, brick is not subject to the rot, wear, and general deterioration problems common to wood. Any repair or replacement will involve the processes

Stair and handrail reconstruction: do's

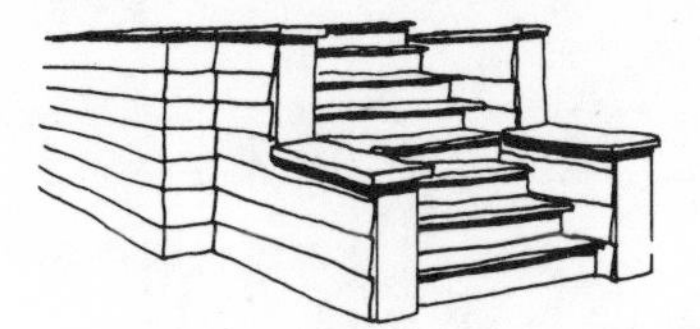

ORIGINAL DESIGN
VISUALLY & PHYSICALLY SOLID, USES CORRECT MATERIAL (WOOD), PROPORTIONS ECHO THOSE OF HOUSE

ADDING HANDRAILS
WHEN REBUILDING STAIRS TO THE ORIGINAL DESIGN, CODE REQUIREMENTS MAY FORCE YOU TO ADD HIGHER HANDRAILS. SOME STYLES OF WOOD STAIRS (COLONIAL REVIVAL, CRAFTSMAN, OR WHEN THERE ARE BUILT-UP SIDES WITHOUT BALUSTERS OR HANDRAILS,) YOU CAN USE INTERIOR BRASS HANDRAIL BRACKETS WITH CLOSET POLES OR 2X2S FOR RAILS.

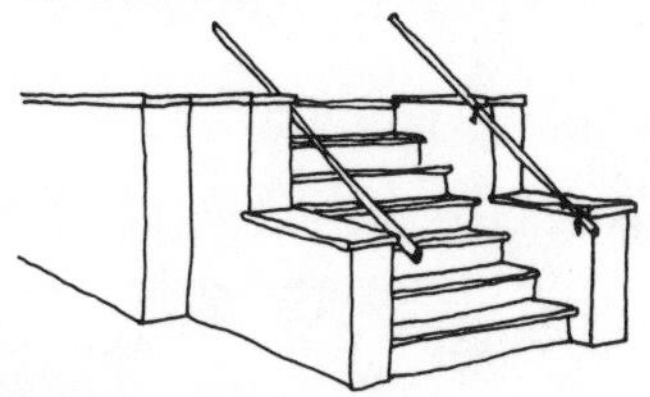

DESIGN INTEGRITY
THINK OF YOUR STAIRCASE AS AN EXTENSION OF THE HOUSE, UNIFIED IN STYLE, RATHER THAN AS A SEPARATE, REPLACEABLE COMPONENT. MAINTAIN OR RESTORE THE ORIGINAL DESIGN & AVOID THE TEMPTATIONS OF READY MADE WROUGHT IRON RAILINGS OR OVERSIMPLIFIED CONSTRUCTION TECHNIQUES.

Porch railings to avoid

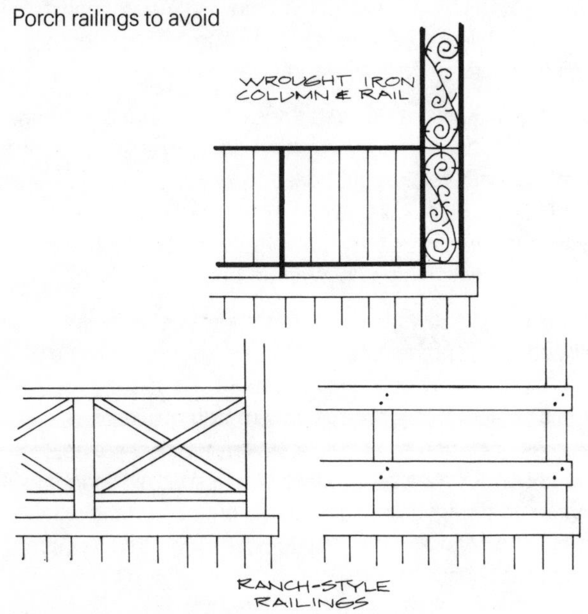

mentioned in Chapter 3, in the section on brick foundations.

When replacing exterior handrails on Victorian homes, *never* substitute wrought iron railings, which are light and flimsy looking and thoroughly out of context in relation to style. Remember, the staircase is an architectural introduction to the home as well as an extension of the house's shape, form, and personality.

Stair and handrail reconstruction: dont's

INAPPROPRIATE DESIGNS

ANGLED SOLID RAILING
PROBLEMS: DOESN'T FIT SQUARE FEATURES OF STYLE, MAKES STAIRS SEEM NARROW & ENCLOSED.

RANCH STYLE RAILING
PROBLEMS: "BACK STAIRS CHARACTER," INSUBSTANTIAL QUALITY, NO RELATIONSHIP TO HOUSE DESIGN.

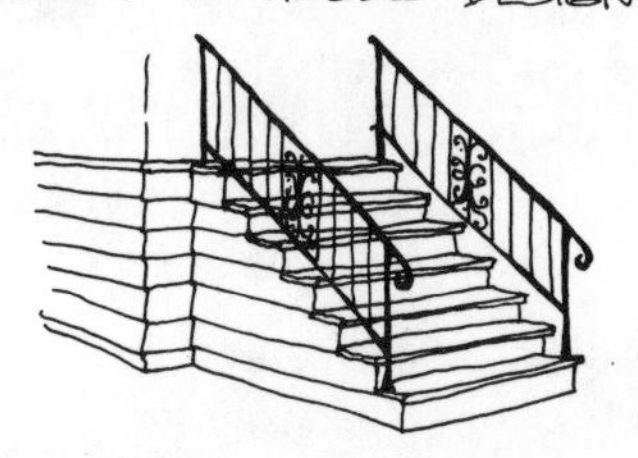

WROUGHT IRON RAILING
PROBLEMS: FLIMSY APPEARANCE, WRONG MATERIAL, INCONGRUOUS CURVED "SPANISH" ORNAMENTS.

BALUSTERS

Balusters are the upright supports to a handrail. If you are going to replace the balusters, you should consider the shoerail approach used for wood railing, which solves drainage problems and repeats the exact proportions of the balustrade elements. Changing these proportions can upset and distort the original design, so always duplicate the original. However, depending upon the specific design, this can be easy or hard.

If the balustrade is composed of flat boards (balusters), use one of the boards to trace a pattern onto a new board. Cut out the pattern with a jigsaw, sand down the rough edges, and place the board into position. Sometimes balusters are constructed in mortise and tenon fashion. That is, each piece has little square appendages that protrude from the top and bottom; the appendages fit into notches in the rails, eliminating the need for toenailing or end-nailing. For mortise and tenon balusters, rout or rerout the notches in the top of the shoerail and the bottom of the handrail. First construct and prepare the shoerail, then do the handrail. Next, squeeze some glue into the notches, and set the new balusters in place. Finally, place the handrail on top of the balusters.

Note: It is often rather difficult to rout out notch holes precisely aligned with the tenon projection. A good

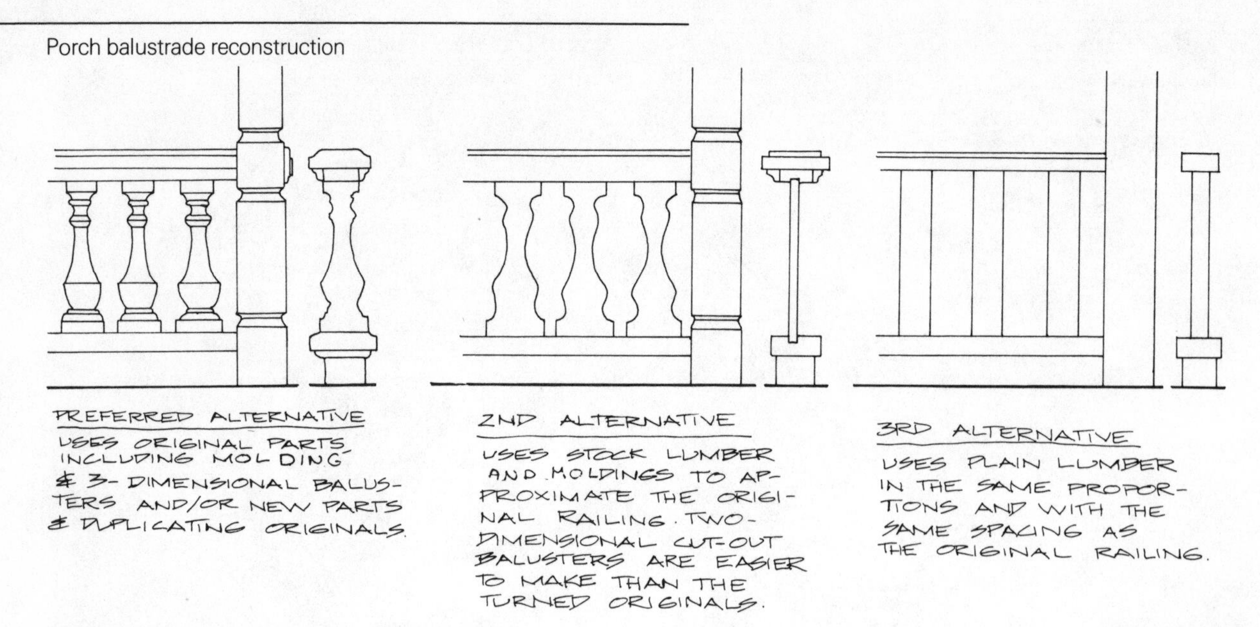

PREFERRED ALTERNATIVE
USES ORIGINAL PARTS INCLUDING MOLDING & 3-DIMENSIONAL BALUSTERS AND/OR NEW PARTS & DUPLICATING ORIGINALS
2ND ALTERNATIVE
USES STOCK LUMBER AND MOLDINGS TO APPROXIMATE THE ORIGINAL RAILING. TWO-DIMENSIONAL CUT-OUT BALUSTERS ARE EASIER TO MAKE THAN THE TURNED ORIGINALS.
3RD ALTERNATIVE
USES PLAIN LUMBER IN THE SAME PROPORTIONS AND WITH THE SAME SPACING AS THE ORIGINAL RAILING.

Porch balustrade reconstruction

remedy and frustration-saver is to rout out an entire line along the bottom face of the handrail. The balusters will simply fit into the slot. Secure the boards in place by toenailing in finishing nails, countersinking them as deeply into the slot as possible. Because this is the underface area, the nails will not be seen.

Two other types of balustrade consist of squared-off pieces and spindles. The square balusters do not require any fancy jigsaw cuts and are relatively easy to replace by using the mortise and tenon joint method. Just drill ½- to ¾-inch holes into the rails and baluster ends. Squeeze a generous amount of glue into the holes, and insert a threaded wooden dowel that fits *very snugly.* This achieves virtually the same type of joint as the one you get with flat boards. Make sure the fit is tight or the dowels will pivot about their axes very easily.

You can replace spindle balustrades in the same manner, but spindles are exceedingly difficult to reconstruct from scratch unless you are an expert with a lathe. The best bet is to buy duplicate spindles and replace them in the mortise and tenon method.

(FRONT PORCH)

Whether spacious enough for a chaise lounge and some chairs, or barely big enough to stand on to escape the rain, the front porch is the focal point of the Victorian facade, and it deserves the time and effort necessary for proper repair. The porch is also a microcosm of the design problems found through the entire house, so this section is a handy introduction to other repairs you may have to undertake.

First differentiate between ornament and structure, and then determine the extent of damage to each. Try to retain as many of the original pieces as possible. This saves you the expense of buying new materials and spares the house the degradation of a porch in an inappropriate architectural style. If the original parts are beyond repair, select new materials that are sympathetic to the design of the house, repeating as much of the original concept as possible. Always remember that the front porch, like the entry stairs, is an extension of the house and so should be treated as an integral architectural feature.

WOOD PORCH

Rot Damage. Wood porches were always found on Colonial and Classic Victorians. But over the years, many wood porches disappeared or were replaced with non-wood materials. If a wooden porch is intrinsic to your house's style, by all means rebuild one or repair the original porch if your house still has it.

If the porch still exists, first detect the amount of damage in the wood. Look for charred wood, with splits along the grain or dark veinlike strands, and wood that has splits and flaking paint. All these symptoms indicate wet rot. *Wet rot* is a fungus that attacks saturated timbers and spreads quickly to nearby wet timbers.

If the wood shows thin white strands, has wool-like sheets with spreading tendrils, feels spongy, or has a multitude of tiny open cells, it has contracted dry rot. *Dry rot* is a microscopic fungus transmitted by airborne spores or carried on shoes or clothing. Under suitable conditions of moisture and warmth, especially standing water, the fungi germinate rapidly on the timbers where they land. Wet and dry rot decrease wood's hardness and toughness.

To investigate the extent of the damage, prod the wood with a sharp tool in an inconspicuous spot and note the resistance to marring. If the wood is sound, the prodding will loosen one or two relatively long slivers, and the breaks will be splintery. Pry out a sliver with a screw-driver. If toughness has been greatly reduced by decay, the wood breaks squarely across the grain and lifts out easily. If the wood is still tough, it splinters and resists removal.

On the porch, look for symptoms of decay on the step treads or deck surfaces, especially those that are checked or concavely worn so they trap water. Decay may also appear at joints in railings; at the base of the posts, especially if the posts are not raised above the porch floor; on the underside of the deck and framing; and at the junction of the porch floor and house wall.

Insect Damage. To detect the presence in wood of harmful insects, look at all areas close to the ground for earthen tubes built over the surface of the foundation and runways from the soil to the wood above. Check for swarming winged adults in the spring and fall, and galleries that follow the grain of the wood. These are all signs of subterranean termites. The insects must return to the ground for moisture, so chemically treating the soil is one line of attack.

Signs of nonsubterranean termites are sandlike pellets discarded outside the wood, swarming winged insects, and tunnels cut freely across the grain of the wood. These pests are most prevalent in the lower latitudes of the United States and are rarely found above 35 degrees north.

Powderpost beetles are indicated by holes the size of pencil lead and borings with the consistency of flour near the ground. Carpenter ants leave piles of coarse sawdust.

The safest approach to dealing with these pests is to call in a reputable exterminator to determine the precise nature and extent of damage. Always request an itemized termite report that defines your home's particular problems. This is helpful as documented proof of the damaged areas; when you proceed to repair insect-damaged wood you can check off the areas as you go along and ensure that every step is taken care of.

Repair. The following three repair steps will return the porch to a safe state and attractive appearance. These are simple or time consuming, depending upon the design of the porch and the extent of the wood deterioration:

1. Remove the ornamentation (see details in Chapter 9).

2. Repair or rebuild the porch structure, which consists of the support posts and piers, the floor joints and decking, the columns that support the porch roof, the safety railings, and the porch roof.

3. Once the structure is repaired or rebuilt, repair and install the ornamentation as instructed in Chapter 9.

Posts and Piers. Replace decayed posts and piers by using a method not susceptible to moisture accumulation. (See Chapter 3, the Post and Pier Foundation, for more instructions.)

Joists and Decking. For drainage, a porch floor should slope slightly downward from the house wall to the railing. However, often the tilt becomes exaggerated because of settlement. To correct this situation, first examine the posts and piers. Be sure they are firmly anchored, intact, and vertical, Next, inspect the joists; if they show no severe rot or termite damage, most likely the problem is merely accelerated settlement. The solution is to raise the floor.

The easiest way to raise the floor is to block up the posts. The procedure is the same one used in replacing a post. First, detach the posts from the girder. Jack up the main joist or girder supported by the posts to relieve the weight. Now, instead of stopping at this point, continue elevating the deck to the desired height. (More than one jack will be needed to ensure uniform weight distribution while elevating the deck.) Next, attach a vertical extension to the top of each post. Be sure the new attachment is the same size lumber as the original and of the correct length. Bolt the end-to-end pieces together with a ¼-inch-thick steel collar or ¼-inch steel strapping. This is critical because the steel straps or collar plates act as the lateral strength of the post at this joint. Now let down the jacks and resecure the girder to the posts. Check the joists; resecure them where needed – they may have been dislodged a bit during the lifting.

If joists need replacing, the decking above probably does too. Remove the decking first. If you need new decking because of wear, consider turning the boards over, which will save the style as well as your money. If you need new decking because of rot, try to replace it with the original type. Sometimes this is an expensive solution because often the original lumber was redwood tongue and groove. However, this lumber is stronger and less likely to warp than conventional boards.

After repairing or replacing the decking, pull out the joist and replace it with a new one. Be sure any new joists are the same size as the original, of redwood, and pretreated with a wood preservative. You can use joist hangers when fitting in the new boards to make the work a lot easier and faster. To retard any potential rotting, treat the joists and underside of the deck with a sealer. If you can stand the smell for about three days, creosote superbly resists the elements and prevents a termite invasion. However, do not use creosote on the surface, because it does not take paint and its murky brown color is ugly. If you consider sealing the surface, use the safest sealer, which is one that is clear and paintable. You could try Cuprenol 20, which is copper green but takes paint.

Columns. Wood columns were used on Victorian and Colonial and Revival houses as structural members or as ornamentation in the half column or pilaster on the facade. The columns are of a classical style, with a base, shaft, and capital. The routed vertical scores on the shaft are called *flutes.* A fluted shaft is hollow, constructed in eight parts like an octagon. A *capital* is shaped like a collar, to fit around the indented top of the shaft. Original capitals are made of exterior plaster, cheap cast iron, or pressed sawdust. Once the seal between the capital and shaft is broken, water penetrates the joint and the capital falls off. If the capital is still attached to your column, be sure the seal is caulked. If the capital is missing, replace it with a cast plaster replica. Local manufacturers in various areas sell these elements.

If an entire column needs to be replaced, buy one from a manufacturer. Fabricating a column takes a great deal of time and requires certain tools and expensive special skills.

Wood columns can be repaired like a porch roof member, in the same manner you would repair the columns on a concrete porch, or porch or foundation posts. If columns that need to be replaced are round or ornamental, buy replicas; if they are simple in design, reconstruct them as you would a handrail (refer to earlier section in this chapter). Be sure the new columns are the proper height, to compensate for any sags or settlement problems.

The wood porches on some Classic Victorians have stone columns, although occasionally the stone is only a veneer over a wooden post. Chemical impurities in the air can decay softer stones, which is evidenced by flaking and pitting in the stone's surface. Treat the stone with a chemical preservative (available from a brick and stone dealer). If decay has completely destroyed the stone, it must be chipped out and replaced. Do your best to replace it with a stone that matches the rest of the column. To protect stone surfaces from future chemical erosion, scrub them regularly with plain cold water. If the mortar binding the stone is crumbling, restore its strength by using the technique described in Chapter 3 for repointing mortar on brick foundations. Finish the mortar with a concave joint, making sure the stone still projects beyond the mortar. Stone columns are important to the architectural character of Classic Victorians and should always be preserved.

Guardrail. A porch railing supports the weight of anyone leaning against it. It is also a visual work of ornamentation, so its particular styling is an important aspect of the overall appearance of the house.

Guardrails incur just about every problem handrails do. According to present-day building codes, if the porch is 2½ feet or more above grade, the guardrail must stand 42 inches above the finished deck surface, with the separation between balusters no greater than 9 inches. Most Victorian, Colonial Revival, and a few other turn-of-the-century styles have rails only 24 or 30 inches high. Your local building ordinances may permit this historical characteristic to be maintained, so you can reconstruct the original design. If you want to increase the railing height for safety purposes, consider the suggestions mentioned in the section on handrails. An alternative is to add a simple raised rail on top of the more elaborate original one. An interior 2-inch-diameter oak handrail intermediately supported by nonferrous metal vertical braces performs the job quite adequately. A milled and painted 2 x 2-inch piece will also serve the same purpose.

Rebuilding ornamental railings from scratch can be difficult and costly. As mentioned for handrails, spindle (or turned) balustrading is the hardest to fabricate. Buy preturned pieces and install them. If this is too costly for your budget, make a silhouette pattern of the piece, trace it onto a flat board that is, say, ⅜- to ½-inch thick, and cut out the shape of the original baluster. Install the flat balusters between the top and bottom horizontals in

Awning substitute for porch

place of the original balusters. The visual effect of style and proportion will make a sincere and sensitive statement. Or you can replace the balusters with simple lumber, say, 2 x 2s, or 2 x 4s, or round pieces. Although this may be technically incorrect as far as the detail of the style is concerned, it makes no emphatic imposition upon the other elements of the facade.

Never replace a guardrail and balustrade with one whose style, character and detail distinctly belongs on another type of architecture. Victorians should never be seen with Spanish wrought iron, ranch-style, or contemporary railings.

Porch Roofs. The two basic roof styles common to the Victorian and Colonial Revival porch are gabled and flat. Both are subject to their own unique problems as well as those common to all roofs. For more information as to solving and repairing these problems, refer to Chapter 4, concerning the house roof; both porch and house roofs are repaired in like manner.

If you want to replace the porch roof at a later date but need a temporary overhead shield, consider a canvas awning, which is functional and also lends a friendly, lyrical appeal to the facade. Manufacturers and suppliers of canvas products offer a variety of designs and colors in addition to the framing, fastening ropes, and hardware; and often they will fabricate the awning to your specifications. Be sure to choose a style reminiscent of the original wooden roof and sympathetic to the proportion, detail and color scheme of the house and neighborhood. And remember that a temporary overhang is just that. Avoid the more durable metal or fiberglas shelters, which require sturdier support and defeat the purpose of being quick and easy short-term substitutes. Fabric has a softer texture and is more readily assimilated into a wood porch.

The roof of some Victorian porches is actually a trellis that provides an intriguing shadow pattern on the porch floor instead of a continuous shelter overhead. Every trellis varies. Look carefully at the lumber sizes, saw-cut details on the ends, and construction joints, and re-create the trellis in the same fashion. This is usually not difficult to do because the trellis has an unpretentious design.

The problems with the trellis are usually its inability to stand squarely upright or the condition of the wood. To firm up a leaning or lopsided trellis, first examine the vertical members. Firmly anchor them onto a steel elevated post base to ward off any future decay. Next, check the joints where the horizontal cross boards and posts meet. Tighten or replace unsecure bolts or pegs with new hardware. Dye or paint new hardware to eliminate its fresh new "tin sequin" look; otherwise the galvanized glimmer in the sunlight will obnoxiously intrude upon the warm, traditional character of the porch.

If the trellis boards need replacing because of decay, substitute with only redwood. Other soft woods are less expensive, but redwood is far more resistant to the elements. Duplicate the saw-cut design in the ends of the boards with a handsaw. If only the ends of the boards are decayed, remove the rotted end (see ilustration) in an L-shaped fashion, creating one-half of a common lap joint. Cut a duplicate of the decayed end portion from a new piece of wood to establish the other half of the lap joint. Place the two pieces together and drill pilot holes completely through the new piece to the old piece, about one-fourth of the way from its midline to the bottom edge. Next, drill smaller, countersunk holes, about half a screw length, into the old wood. Be sure the screw is of such a length that when it is fastened it comes no closer than 1 inch from the bottom of the old board. Do *not* let the screw poke through. Tighten the screw, squeeze a generous amount of wood glue into the pilot holes, and cap the holes with a snugly fitting wooden dowel. Let the glue dry.

The main problems with this economical solution are matching the size of the boards and disguising the joint. And even if you manage to accomplish this successfully, the seam will still tend to collect moisture, starting the decaying process all over. You may want to replace the entire piece instead.

Many trellises are covered with wisteria or some other type of crawling vine. As fresh and welcoming as this looks, especially when the wisteria is in bloom, plant growth on wood kills its molecular structure because the vegetation attaches itself to the wood and nurses on the nutrients the wood provides. The strength of the wood is thus weakened, and the decaying process is greatly accelerated. But because foliage on a trellis is often a design feature of the entry, it is really a shame to destroy it. A good solution is to delicately detach and lift the vines off the wood (as you would if you were repairing the trellis), treat the wood with a preservative or paint it, and then replace the foliage in its former position. Usually it will automatically resume its clinging and coiling growth process.

STUCCO PORCH

The stucco porch is found on some Victorian houses, and it is also found on houses that initially had wood porches but have since been modernized. If your stucco porch needs repair and stucco was not the original material, consider replacing the porch with one better suited to the architecture, which in most cases will be a wood porch.

Stucco is a many-layered material. The structural wood frame is covered by tar paper, the tar paper by wire or wood lath, and the lath by three coats of a stucco mixture, which is primarily Portland cement and sand. Any or all of these layers may need repairing.

Repairing small cracks or holes in the stucco surface is not difficult, but matching the texture and getting the stucco to adhere so it does not look like an obvious patch job is definitely time consuming. A professional can do the job in a relatively short time, whereas it may get on your nerves. (If you do stucco work yourself, refer to Chapter 5, the section on stucco.) You must keep cracks patched or they will enlarge and allow moisture to seep into the structure underneath and cause deterioration that can go unnoticed for quite some time. When the damage is finally detected, you might have some hefty and extensive repair work to do.

Cracks and holes are mainly caused by the underlying wood framework because wood is subject to decay from moisture accumulation, poor air circulation, and termite damage. If you recently bought your home and have a termite report, your porch may be listed. It may be difficult for you to notice the damage on a stucco structure, but if it has been detected by a professional, you had better take a closer look. Look for cracks, faulty gutters, and inadequate flashing, which permit misdirected water runoff or leaks in a flat tar and gravel roof. If the damage is severe, the entire porch must be rebuilt and restuccoed. Seek the assistance of a contractor.

Another problem with stucco porches is rot in the joints between the porch floor and the framework or house because of improper or omitted flashing. *Flashing* is a coved strip of building paper or galvanized metal that fits over the joints or seams. It diverts water from places where it can collect. If there is no flashing at these critical junctions, water will be absorbed into the seams and deteriorate the stucco and the cement-covered wood decking. Correcting this situation requires removing the damaged materials prior to installing the flashing. Fit the flashing in firmly and flat enough so as not to produce an obvious bulging line on the floor and wall surfaces when you recoat the stucco.

Flashing is equally important at the joint between the porch roof and wall of the house. If water is allowed to accumulate here, it can penetrate to the wood frame and cause serious structural damage. (See Chapter 4, the section on flashing.)

A final problem with stucco porches is one of function rather than maintenance. Many people, seeking extra living space, enclose the front porch. But all this does is create a very small and drafty room. The gracious introduction to the home as well as the entry platform and rain shelter are sacrificed in the swap. Besides, building codes require that a 3 x 3-foot minimum floor platform be directly adjacent to entry/exit ways—the logic being that an exterior floor surface provides a safer means of exit than does the arrangement of an exit door and then steps. If your porch is still a porch, do *not* enclose it.

If you absolutely need a windbreak or sunscreen to make the porch more comfortable for sitting or growing plants, use plain transparent plastic or glass. This will accomplish your goal without disrupting the traditional appearance of the porch.

CONCRETE PORCH

The concrete slab porch is common to the Mission Revival style of the 1920s. And although not authentic to turn-of-the-century homes, concrete porches are presently found on many of them. Wood porches and exterior steps are highly vulnerable to decay, so they are often the first two features to be replaced, frequently with more durable concrete.

There are two opposing views about a concrete porch that has replaced a wooden one: (1) it is historically incorrect and does not belong to the architecture; or (2) there are now so many of these replacements that they have become an accepted style in themselves.

If you hold to the latter view, maintain and repair your concrete porch where and when necessary. Look for cracks and signs of uneven settlement, as you would in masonry stairs and foundations, and repair them with the same techniques. However, beware of a network of cracks, a split that horizontally shears the slab into two pieces, or a cracked slab with a shift in elevation from one piece to the other. These can be signs of severe soil instability or a tree root bulging up from below. Consult an engineer to determine the specific nature of the problem.

Make sure the slab does not slope toward the house. Place a marble or ball on the slope and see which way it

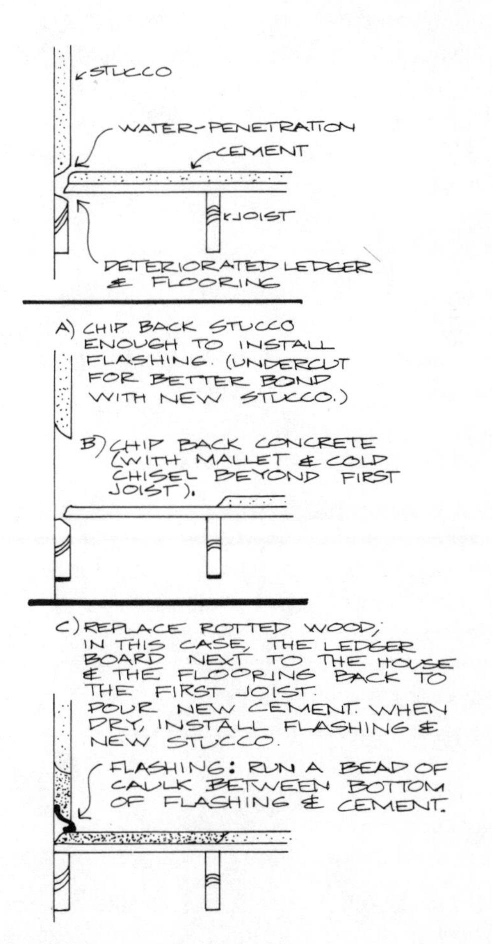

Concrete porch floor repair with flashing

rolls. If it heads for the door or walls, water must also drain in that direction. If the wood siding is decaying at the porch and wall junction, the problem needs immediate attention because there is likely to be hidden rot behind the siding and in the floor joists.

If you feel that concrete porches do not belong on historical buildings, you are morally and technically justified. However, demolishing a concrete porch and replacing it with a wood one can be quite a chore and expensive. A decent compromise is to leave the masonry construction, repairing it as necessary, but placing authentic wood decking over it. But be sure drainage will not be a problem. In addition to ensuring the proper slope of the porch and steps, raise the deck ⅜ to ½ inch up off of the concrete surface and anchor creosote-treated strips to the concrete with ¼-inch-diameter steel expansion bolts. Place wooden strips perpendicular to the house or in the direction of the slope, spaced 12 to 16 inches apart. Also place strips on the stair treads and risers, but lessen their length. Nail the decking across the strips.

Tongue and groove that is properly sealed works the best for the stairs. Use the appropriate sizes of boards. A telltale sign will be visible where the tread and riser meet if you do not extend the riser up high enough to cover the creosoted strips. Be sure the gap is closed and that the tread nosing projects ¾ to 1¼ inch over the riser. If the threshold of the front door is high enough, there should be no problems with an additional deck elevation of 2 inches or so. If the new deck surface will be level with or higher than the threshold, do not use this alternative.

(CHAPTER 9)

ORNAMENTATION AND PAINTING

Often it is the decorative detail of trimwork that defines the essence of an architectural style. When describing or determining design and period, for instance, you will probably first distinguish the particular types of motifs and ornamentations that embellish the structure. This is why the preservation of ornamentation is essential for maintaining the proper antiquity of the house.

Note that many West Coast towns and cities have passed ordinances that prohibit certain concrete or plaster overhangs and cornices from projecting over inhabited areas becaue of possible earthquake damage. Nonreinforced brittle material, such as plaster or masonry, tends to break off and crumble during a quake, and it may cause damage and kill or injure people when it falls. Before reconstructing any elaborate details, be sure you get local approval.

(WOOD ORNAMENTATION)

REMOVAL

It is often necessary to remove ornamentation to repair the underlying structure or to repair the ornamentation itself. Do this with extreme care and patience. Replacing certain types of decorative detail *you* have damaged is not only costly, but you may not be able to duplicate the design if it is rare. Study the part first to see if it is attached with a toenail or facenail. Use the broad surface of a prying instrument like a crowbar, working it very gradually to loosen the knobs from the supporting posts and beams. When space allows, place a piece of corrugated cardboard or wood between the tool and the structural wood to avoid denting the surface as you unnail.

A cedar shingle is a terrific tool for removing ornamentation. It can be split lengthwise so its width is the same as the part in question, and pressure can be distributed evenly. It is thin enough because its tapered end allows for gradual insertion, and since it is wood it will not scar the ornament the way a metal tool will. The shingle also has a fail-safe mechanism: when it bumps into a nail, it automatically splits.

If paint masks the joint between the ornament and its backing, score the seal with a putty knife. Then gently tap the tapered end of a cedar shingle into a crack on an unnailed edge until the force of the shingle lifts the ornament away from its support. Ease the shingle out, keeping it horizontal. Do not use the shingle as a prying mechanism or it will break off. There should now be enough space to permit purchase on the nail with a claw hammer crowbar. (Refer to Chapter 11, the Trim section, for more removal techniques.)

With an indelible marker, label the pieces of ornamentation to help you when you reassemble them later. You could also draw a sketch as a convenient record, or take a "before" photograph. Store the collection of ornamental parts in a weather-protected place, inaccessible to vandals.

REPAIR

If ornamentation is beyond repair, save the pieces and use them as a pattern. Even if the intricate detail is lost in

reproduction, combining standard geometric parts in the proper proportion suffices well as long as the visual relationship among the parts remains the same. For example, perpetuate the appearance of a corner board by using a plain 2 × 8 inch, or keep window trim broad enough by using a 2 × 2 inch for molding. A rather proportional reproduction with its corresponding re-creation of a cornice is illustrated.

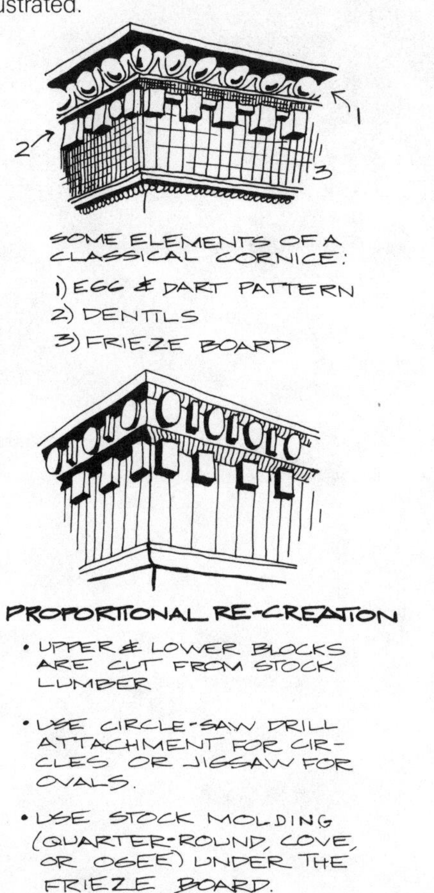

1
2
3
SOME ELEMENTS OF A CLASSICAL CORNICE:
1) EGG & DART PATTERN
2) DENTILS
3) FRIEZE BOARD
PROPORTIONAL RE-CREATION
• UPPER & LOWER BLOCKS ARE CUT FROM STOCK LUMBER
• USE CIRCLE-SAW DRILL ATTACHMENT FOR CIRCLES OR JIGSAW FOR OVALS.
• USE STOCK MOLDING (QUARTER-ROUND, COVE, OR OGEE) UNDER THE FRIEZE BOARD.

Proportional reproduction — a cornice

If you prefer accurate duplication, bring the part to an historical house-part supplier, who may have the period detail in stock. If not, the reproduction can be custom made by a supplier. However, this service can really cost if you are repairing a large quantity of ornamentation, such as the entire perimeter of the house, trim from every window and every door, and so on. Carefully consider costs, and remember that exterior detail does not usually receive the same close-range scrutiny as does some of the eye-level interior detail; so other areas of repair may be more worthy of financial attention.

You could also make a latex cast of the original detail; a reasonable facsimile can then be molded from plaster. (See the instructions for making a cast from a plaster original in the Plaster Ornamentation section.)

Finally, you can rummage through salvage yards to see if they have the authentic pieces you're looking for. Just make sure that whatever you acquire matches whatever you already have, or an inappropriate mixture of detail will look like a gaudy hodgepodge. Avoid plastic reproductions that do not carry the tone, texture, proportion, and proper detail of the original molding. When applied en masse, plastic decoration can look like tacky Christmas ornaments.

REINSTALLATION

Correct installation techniques always apply. When reinstalling any type of trimwork, pay attention to nailing. If you use flat nails, be sure they are concealed by other molding. If you cannot cover nails with overlapping boards, countersink finishing nails and fill in the holes with patch. As the trim begins to integrate itself with the rest of the structure, it will show signs of irregular settlement and may lift, bend, or pop off of the facade. The nails may begin to jar out of or through the molding. Countersinking lessens the potential disruption of the smooth, even surface.

Treat all parts, both old and new, with a wood preservative. By doing this before you reassemble, all parts receive equal protection. You can be sloppy during this process. Spread out the pieces on a plastic drop cloth and brush them well. When they are thoroughly dry and the underlying structure has been repaired, nail the pieces of ornamentation back on.

Sometimes the underlying structure is in good shape but far from flush, plumb, and square. Flat ornamentation applied to a somewhat buckled surface can pose a problem. If the wall is warped considerably, shim it flush to the molding with a wooden wedge. Then nail or screw the molding tightly in place. If you are attaching old wood with screws, remember to drill pilot holes first because the wood may be brittle and tend to crack. Always caulk the molding joints and keep them protected with a decent coat of paint.

[PLASTER ORNAMENTATION]

The yen for structural ornamentation during Victorian and Colonial Revival periods was often satisfied with plaster instead of wood. On the exterior, plaster was used for capitals on classic columns; as rosettes, festoons, and garlands at the cornice; and for obscure accessories on eclectic facades.

Plaster ornamentation is often deteriorated or missing because the material itself weathered or the attachment failed and the decorative piece fell off. If you have at least one decorative piece of plaster ornamentation left, remove it carefully and make a mold from it, to cast replacements. To make a mold, use a commercial product such as Liquid Latex. Brush on continuous coats and let each one dry. They form a rubbery mold, which you remove and cast in an exterior plaster. If you plan to mold the ornament onto the facade, embed an anchor first, as you would in a foundation.

If there is no original plaster work left, you can buy replacements, but be sure the plaster ornaments *belong* on the house; research the proper motif before reinstalling any plaster elements. Historical supply shops often stock many designs. Remember to keep the replacements free of moisture damage by caulking and sealing any joints between the plaster and structure, to prevent water from seeping into the joints and dislodging the piece altogether.

(PAINT)

A good coat of paint is one of the best defenses a house has against the elements. Paint forms a continuous film that sheds water, a primary source of building damage. Paint is also one of the most important visual selections you will be making during the restoration process. The color combination decides the overall appearance of the house and contributes to the character of the entire neighborhood.

WHEN TO PAINT

Trouble signs that indicate a new paint job is necessary are alligatoring (scaling), checking, cracking, sealing, blistering, peeling, and exposed wood. Paint that rubs off like powder is quite normal; good outdoor paint is designed to chalk so that rain will wash away dirt and leave a clean surface. If the paint is dirty or faded, it still should provide a protection seal for the wood or masonry underneath, but it can be freshened up by a mild detergent. Trim is more vulnerable to weathering than the body of the house, so if paint shows signs of wear only there, limit the job to repainting the trim. Do not rush to entirely repaint the house, because excessive coats create a thick film that can promote damage. Do not undertake a paint job until any problems with leaking water have been solved, especially the repair of gutters and downspouts.

Many people prefer to leave the paint job to the last, as the crowning touch to the restoration process, the climax to the entire operation. Other people paint early, finding it easier to get loans and insurance if the house looks presentable. Still other people paint the house just to make it look attractive to prospective buyers. As effective as this sales pitch may be, essential underlying repair work is often neglected. Thus the paint job is really a waste of time and energy. And this ploy is unfair to new owners whose taste in color may be different than that of the previous owners. Even more important, high-quality paints more than pay for themselves in the long run, but a person about to sell a house has little incentive to use an expensive product. If buyers do the job instead, they will wisely select a top-flight product that will prolong the life of the paint job and make subsequent jobs easier.

THE PAINTING PROCESS

There are four stages to painting the exterior surface of any house: surface preparation, color selection, paint application, and clean up. Surface preparation is at least half the job and worth the extra time because good preparation is the key to the ultimate success and duration of the paint job.

Preparation and Application. Preparation involves:

1. Removing the loose paint with a wire brush, scraper, chemicals, or heat;
2. Fine sanding the scraped surface to feather out rough edges;
3. Lightly sanding the smooth base to give it some tooth;
4. Nailing, puttying, and caulking as needed;
5. Cleaning off all dirt and dust.

The actual paint application involves:

1. Using a primer as needed;
2. Keeping the paint thoroughly mixed;
3. Brushing or rolling on the color in an orderly way.

On single-story houses it is easier to paint the body of the house first and then do the trim. On two- or three-story houses it is easier to complete both base and trim a section at a time. All that is left then is the cleanup. (You can rent scaffolding for $200 to $300 a month if you want to use it for painting a two- or three-story house.)

Color Selection. As a positive influence, color is contagious. It often takes only one house to start the trend

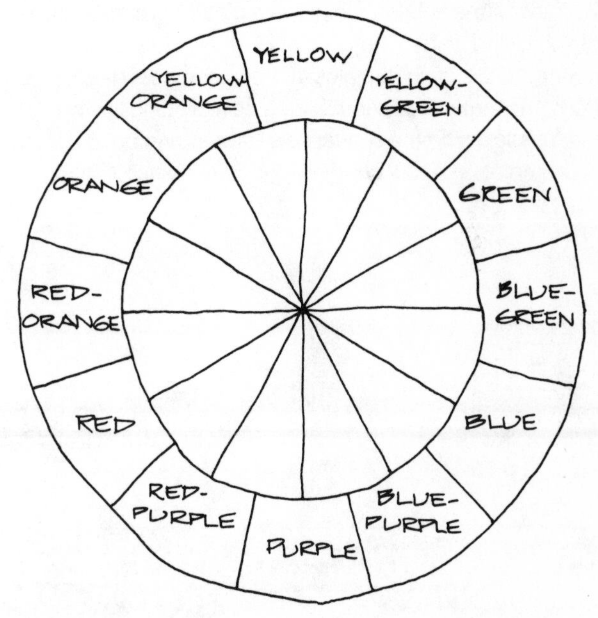

Color wheel

toward a fresh coat for the whole block. But an odd color can be a potential threat in the local environment. Blending or contrasting colors with nearby buildings is as critical as choosing compatible colors for the house itself. In case of any doubt, the best bet is to assume a low profile. For example, if the street is overall pastel, flow with the tide of subdued colors. This is important if adjacent houses share the same architectural style because cacophony is difficult to forgive. And where there is little landscaping to soften masses and few trees to filter the view of the facade, considerate color selection is even more critical.

Use the paint palette to help a house really assert *its* authentic personality rather than your own. As with other aspects of appearance, the historically correct color choice is usually a safe one. If you stray drastically from historic guidelines, you just may downgrade the appeal and resale value of your home. For greater assurance when selecting the precise color for your Victorian, here are some hints.

1. Look at the house as a whole. Study its shape and proportion, the arrangement of the parts and their relationship to one another. Look for geometric shapes like squares, rectangles, triangles, and polygons created by the shadows on the building's facade. Study the components, like the roof, body of the house, and the base level. Analyze what portions are (or should be) emphasized or subdued. To determine prominent versus recessive areas, ask yourself the following questions: Does the roof cloak the building, or is it merely a finishing cap? Are the cornices and trimwork eye-catching features? Do the porch, stairs, and entryway present a separate structural statement of their own? If you can determine what some of these key elements are attempting to say, it will be easier to select appropriate colors.

2. Think of the house's facade as a picture, a composition of solids and voids. The walls appear as light planes punctured by the dark holes of the windows. The change from one material to another creates a dividing line on the facade, as do the sequence of stories and bands of ornamentation. A successful color combination unifies the architectural elements into a single picture, without denying each distinctive feature.

3. Decide how much of the facade is actually paintable. There is probably a lot more window space than you thought. Do *not* paint stained shingles, brick, stone, untreated wrought iron, the wood stickwork on Brown Shingles or Classic Bungalows, the chimney, the roofing, or any modern sidings. *Do* paint the base, the trim, window moldings, the moving parts of wood windows, and the wood or stucco siding of the house.

Shades and Tints. Pure hues (colors) become shades when black is added to them, tints when white is added.

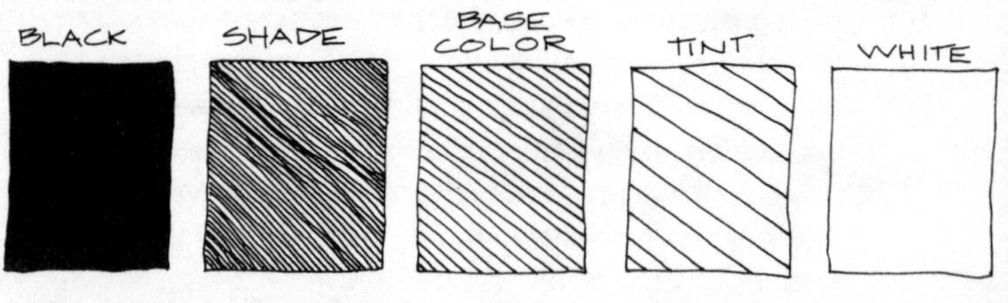

Parts of house to paint

Adding a hue's complement (the color opposite it on the color wheel) will gray or mute the color. Earth tones, such as the various browns, greens, and ochres, are made from muting or mixing complementary colors.

Your safest bet is to stick with shades and tints of raw hues or earth tones. Limit the selection to no more than three. Economy is a consideration, but appearance is the primary concern. The most acceptable rule of thumb to follow is: a *light or muted* color for the *body;* a compatible *darker* shade for the *base, trim,* and *window moldings;* and *white* for the *moving parts of* wood frame *windows.* This gives you a three-color effect (although white is not actually a color). Two other visually intriguing combinations are a muted body with dramatically con-

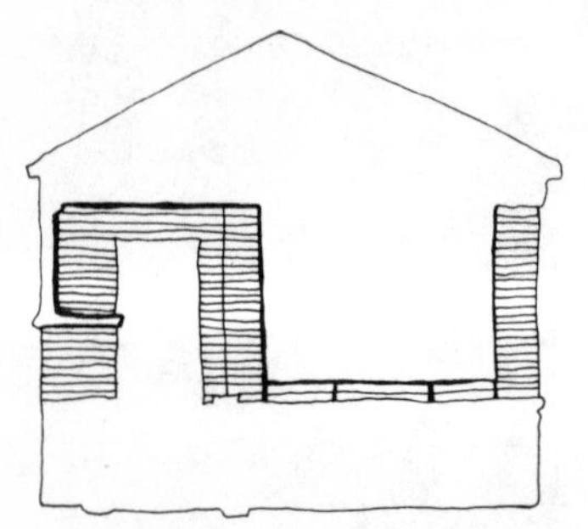

THE BODY: GENERALLY CONSIDERED THE BASIC COLOR OF THE HOUSE, THE BODY MAY HAVE A SURPRISINGLY SMALL SURFACE AREA ON THE STREETSIDE FACADE. IN THE NEOCLASSIC ROW-HOUSE EXAMPLE, WHITE OR A LIGHT COLOR WILL EMPHASIZE THE HORIZONTAL SHADOW PATTERN INTENDED BY THE NARROW SIDING.

THE TRIM: TRIMWORK RECEIVES THE SECOND COLOR IN A TWO-COLOR PAINT SCHEME. OFF-WHITE OR LIGHT-COLORED TRIM GOES WITH THE ATTRACTIVE DEEP EARTH TONES. WHEN LIGHT COLORS ARE USED FOR THE BODY, A DARKER VARIANT OF THE BODY COLOR WORKS WELL FOR TRIM.

OPPORTUNITIES FOR MORE COLOR ARE PRESENTED BY WINDOWS, DOORS AND STAIRS. MOVING WINDOW PARTS ARE BEST PAINTED WHITE, LIGHT, OR BLACK. WHITE IS PLEASANT SEEN FROM THE INTERIOR AS IT REFLECTS LIGHT INSIDE. STAIRS AND PORCHES ARE TREATED WITH SPECIAL HEAVY-DUTY PAINTS, OFTEN IN A NEUTRAL GRAY. DOORS CAN BE A BRIGHT SPOT, BUT UNLESS YOUR HOUSE IS NEUTRAL (GRAY OR WHITE) PICK A COLOR THAT HARMONIZES WITH BODY AND TRIM.

Parts of house to paint

trasting trim and a compatibly lighter tint of the body color for the moving parts of the window; or a deep earth-tone body highlighted with white or very light trim and relatively darker window casings.

For a two-color scheme, paint the base, body, and moving parts of the windows a light or muted color and accent the trim with a compatibly darker shade; or paint the trim a light color and the remainder of the house a darker shade. It is often shameful to drown the rich form, shape, and detail by painting the house a single color; but if economy warrants this, your wisest choice is a light to neutral muted tone. Avoid dark colors.

Avoid loud, raw hues; combinations of primary or complementary colors such as red, green, and yellow; vivid purple; or any color that might make your home look like a playroom or head shop. Remember, you, as well as passersby, are chronically subject to your environment; sometimes colors that catch your fancy one moment may drive you and anyone else crazy if you have to live with it. Some cities and neighborhoods have actually established ordinances that give detailed guidelines as to what exterior colors are and are not permissible. These decisions are deliberately intended to discourage "passion" color selection.

Generally, dark colors make small areas more intense. For expansive walls, deep earth tones are acceptable because they visually reinforce the feeling of structural stability. Dark colors are also used as a field for limited amounts of ornamentation painted off-white.

Light colors do justice to large planes, like the side of a stucco bungalow, because the color reduces the heavy massiveness of the wall. Pastels, tints of pure hues, are popular in Mediterranean and other warm climates because they reflect rather than absorb heat and so are cooler and "airier" to look at. Do not use lavender or aqua pastels because they are historically inappropriate; they are a dated expression of the Eisenhower era. A light color is the safest choice, but it can be the dullest unless you enliven it with exciting trimwork contrasts.

There are as many different whites as there are colors. The one to select for trim should be an extremely light tint of the color used on the body of the house. White is the best selection for formal styles whose walls are visually contained around the edges. The clapboard walls on the Colonial Revival styles are contained by the plasters on the sides and the pediment on the top. The smooth stucco walls on Mediterranean houses are capped by a change in texture: the rough terra cotta of the red tile roof. Both styles look best when painted a crisp white.

Sparkling contrasts of dark and light colors perk up a paint job when used in a limited way. Accentuate window moldings and doors to emphasize the house's strength of character. However, if contrast is misused, parts of the facade will visually separate into distinct, independent rather than unifying elements. If the proportion of dark and light planes is equal, the surfaces seem to advance and recede at will; but if contrast is misapplied to ornamentation, the building can look like it is ribbon wrapped.

A sure way to define the form of projecting parts is to stand back from the house and notice how the light hits it. Direct sun works best to define shadow lines and shapes cast on the building. Shadows are important elements because many house styles, such as those with projecting bay windows, accentuated roof overhangs, or clapboard siding, expressly rely upon light's dimensional quality to create part of the facade's appeal of light and shadow. White is one of the best choices for "getting the most out of" the lines of contrast created by shadow. The white will almost glisten, and the shaded areas will appear stark black. White is especially useful for creating this effect on large boxlike structures like Colonial Revivals.

Avoid dangerous combinations of warm and cold colors, like red and blue or blue-greens and vivid oranges. Two colors of the same intensity are also undesirable because the colors fight each other, each vying for the most attention. Also be aware of architectural features that have their own color such as the roof, chimney, foundation, and any stained glass windows. And the color of adjacent concrete walks and foliage are just as critical to the overall color scheme as those on the house itself. For instance, if your foundation, stairs, and/or walkway are brick, do not antagonize their natural color by painting the house say, olive green and persimmon orange just because you like that combination. As enchanting as these colors may be, modify your personal tastes to integrate with the building's personality. For example, in this case try a terra cotta or even a bright rust instead of the overly warm persimmon. Against a warm beige base, the rich green and red earth tones as trim will not only satisfy a spicy appeal, they will best complement existing structural features.

[PART THREE]

INTERIOR ELEMENTS

In this book we are concerned primarily with the basic interiors of the house — floors, walls, woodwork, ceilings, staircases and doors, rather than the interior space planning. It is essential that the house be sound in all these respects. It is these areas that generally need work — sometimes a great deal of work — but what is involved is nothing more than basic carpentry and repair.

Whether you want the interior to appear as it did when the house was originally built or rather opt for intelligent replacement of parts of the inside materials to your own tastes is your own decision. Here we give information on making your Victorian as sound as possible to weather the evils of time.

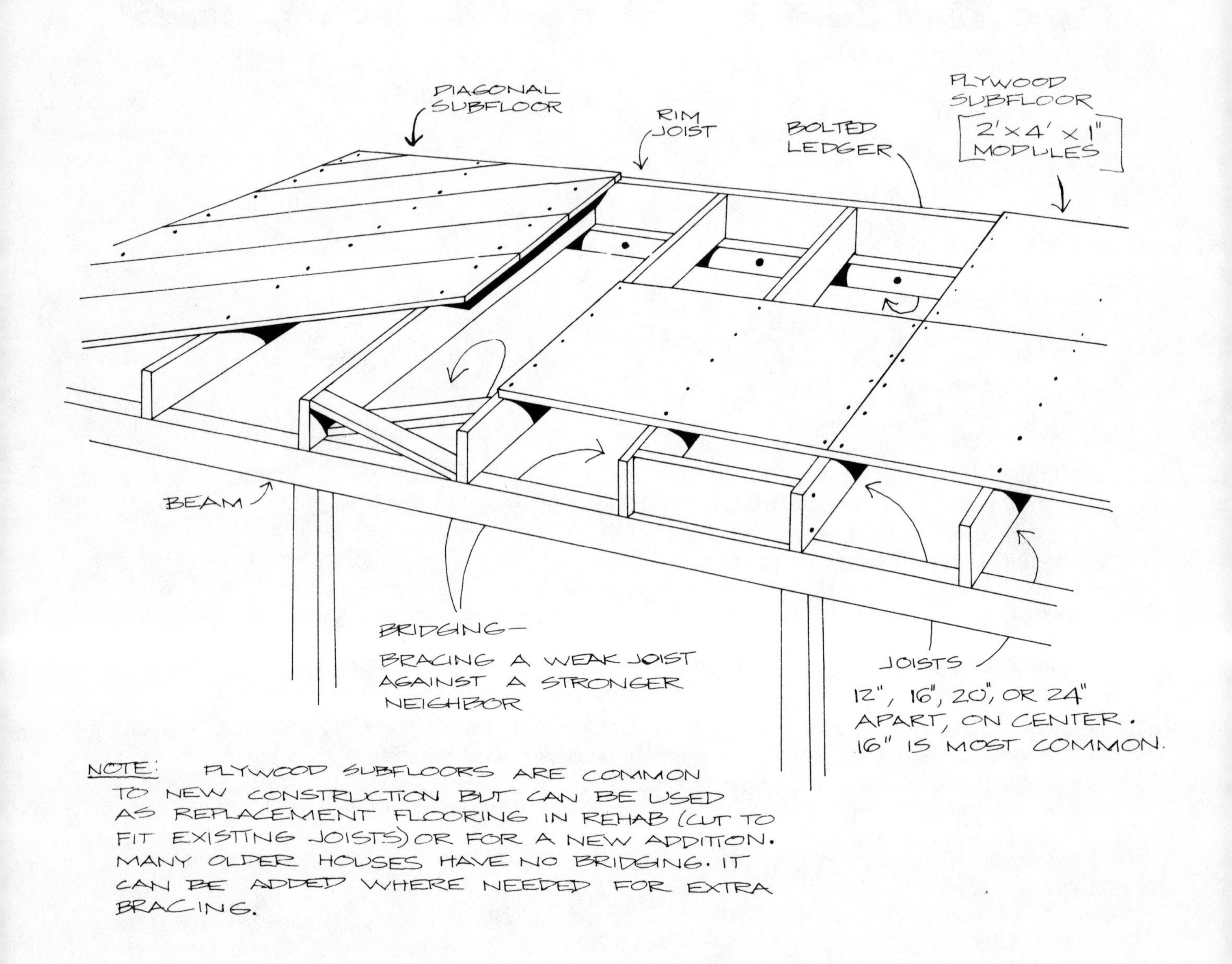
DIAGONAL SUBFLOOR
RIM JOIST
BOLTED LEDGER
PLYWOOD SUBFLOOR
[2' × 4' × 1" MODULES]
BEAM
BRIDGING—
BRACING A WEAK JOIST AGAINST A STRONGER NEIGHBOR
JOISTS
12", 16", 20", OR 24" APART, ON CENTER. 16" IS MOST COMMON.
NOTE: PLYWOOD SUBFLOORS ARE COMMON TO NEW CONSTRUCTION BUT CAN BE USED AS REPLACEMENT FLOORING IN REHAB (CUT TO FIT EXISTING JOISTS) OR FOR A NEW ADDITION. MANY OLDER HOUSES HAVE NO BRIDGING. IT CAN BE ADDED WHERE NEEDED FOR EXTRA BRACING.

Parts of a floor

(CHAPTER 10)

FLOORS

Floors receive an enormous amount of wear and tear. When you set out to improve the floors, make the job as good as possible by using the right materials and careful workmanship.

(SUBFLOOR)

The subfloor is a rough base for the finished floor. It is supported by joists when a concrete slab foundation is not used, as in all prewar houses. The oldest type of subfloor consists of wood planks nailed perpendicular to the joists. Beginning in the 1920s, the boards were nailed at a 45-degree angle to the joists. Since the 1950s, subfloors have been made of plywood. With the diagonal plank and plywood subfloors the finish wood stripping can be run in any desired direction, whereas with perpendicular-to-joist subflooring, the finish stripping had to be laid at right angles.

The subfloor rarely needs attention. However, as wood dries and ages it settles into certain positions, and sometimes the joists may sag and the subfloor start to shrink. Squeaks underfoot denote this loose fit. To correct these minor irritants, first find the source of the squeak. Go underneath the floor, to the joists below the squeak. (To gain access to the subfloor, carefully remove the finish floor and the baseboard.) You may discover the sinister gap that is causing the problem. Tightly wedge some cardboard or soft wood into the space, to compensate for the void and to absorb weight properly, eliminating the squeak.

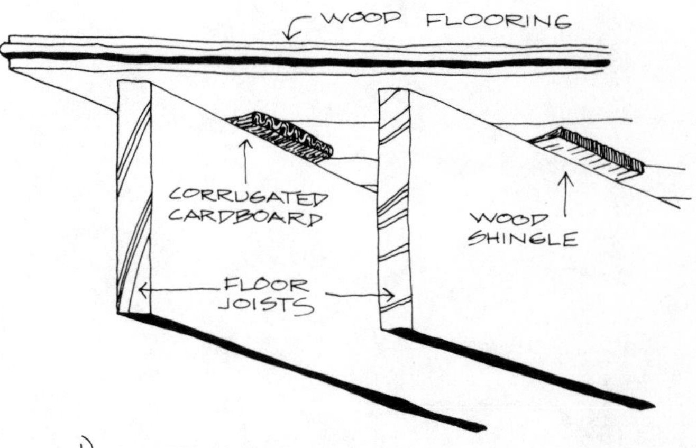

WOOD FLOORING
CORRUGATED CARDBOARD
WOOD SHINGLE
FLOOR JOISTS
1) LOCATE SQUEAK BY STEPPING AROUND ON THE WOOD FLOOR.
2) LOCATE JOIST DIRECTLY "BELOW THE SQUEAK".
3) WEDGE A PIECE OF CORRUGATED CARDBOARD OR A WOOD SHINGLE BETWEEN THE JOIST AND FLOOR WHERE SQUEAKING IS OCCURING.

Fixing squeaky floors

Another problem with subfloors is rot caused by excessive water. This occurs primarily in the kitchen, near the sink, and in the bathroom, near the bathtub and toilet. The finish floor buckles, discolors, or gets a spongy texture.

If you have to replace the subfloor, always locate and correct the cause of the problem beforehand. Sometimes it is difficult to replace a portion of the original subfloor with the identical type and thickness of wood because the original may be of some obscure dimension or of sub-

Replacing damaged floorboards

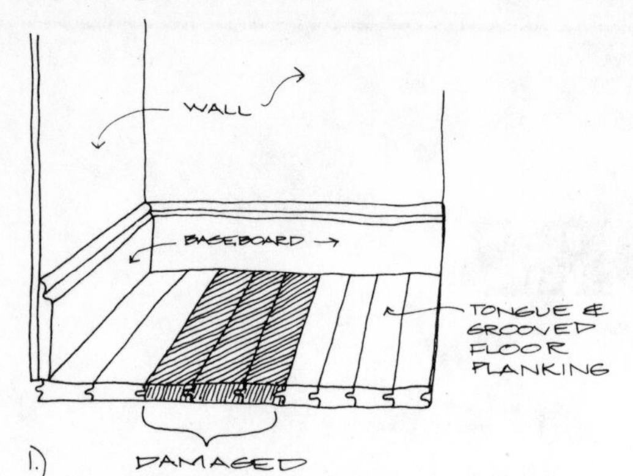

1.) DAMAGED FLOORBOARDS

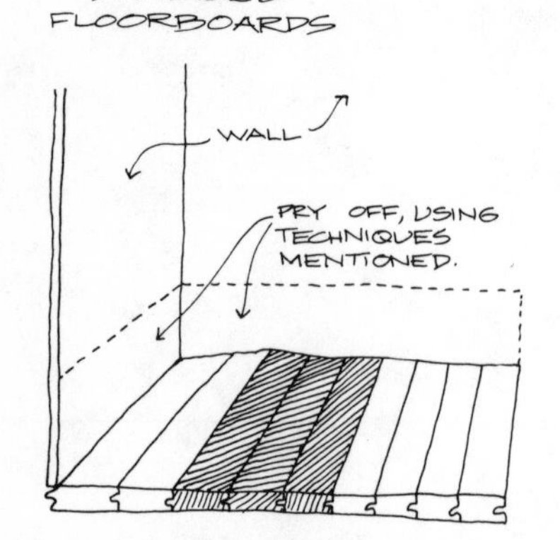

2) REMOVE BASEBOARD. AVOID DISTURBING PLASTER OR GYPSUM IF POSSIBLE.

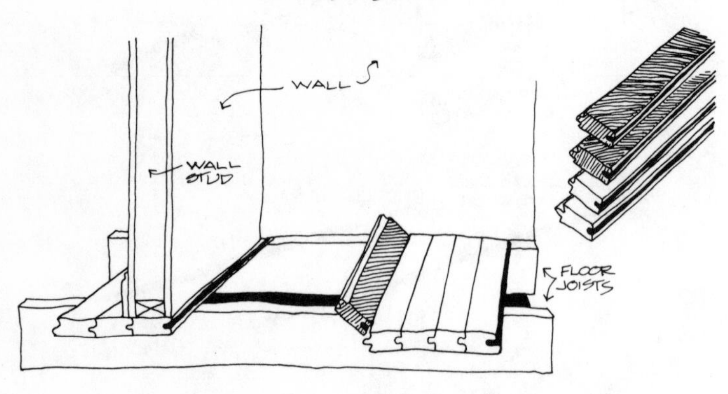

3) LOCATE WALL NEAREST TO DAMAGED FLOORBOARDS. WITH A CROWBAR, PRY UP EACH BOARD IN SUCCESSION. STARTING FROM THE PLANK NEAREST THE WALL, CONTINUE TO LIFT UP THE BOARDS UNTIL THE DAMAGED PIECES ARE REMOVED.

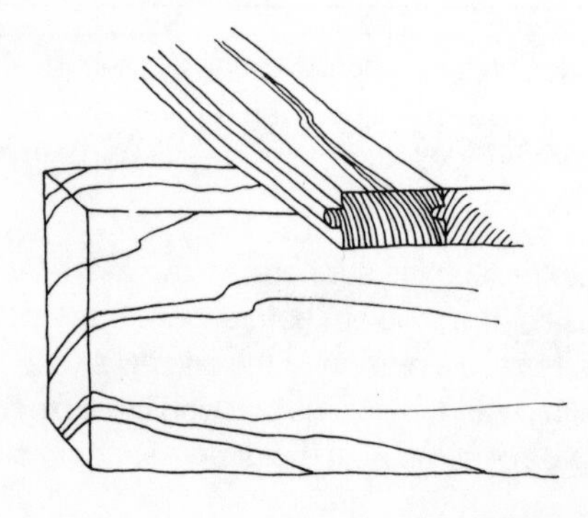

4) REPLACE BOARDS IN-KIND, NAILING EACH SUCCESSIVE BOARD TO JOIST. START AT END FARTHEST FROM WALL & WORK BACK TO WALL.

5) REPLACE BASEBOARD.

6) FINISH BOARDS TO MATCH THE EXISTING FLOOR.

NOTE: REPLACEMENT BOARDS MAY BE OBTAINED FROM CLOSET FLOORS OR OTHER "INSIGNIFICANT" AREAS.

standard thickness according to current code. (Some ordinances require ½- or ⅝-inch subflooring, depending upon the type of dwelling.) If your rot damage is minor, try your best to match the original subflooring, but if the damage is extensive, replace the subfloor of the entire room with the properly dimensioned plywood.

Plywood is the best substitute for the original subflooring because it does not involve much labor, gives you freedom of choice for finish floor installation, and provides better resistance to horizontal shift from settlement. Firmly nail the plywood to the joists, and tightly abut the pieces to prevent air gaps and spider hangouts.

[FINISH FLOOR]

The finish floor is the floor you walk on. Unlike the rough subfloor concealed beneath it, the finish floor is a formal part of the room, so the original flooring was selected to match the interior space. Many Victorians did not have finish floors because wall-to-wall carpeting was installed directly over the subfloor. Beginning in the 1890s, the finish floor was made of standardized wood strips, odd-lot planks, or checkerboard parquets. Hardwood was typically used for living and dining rooms; softwood was usually reserved for pantries and the second story.

HARDWOOD

A hardwood floor is one of a home's most valuable resale items. Check the newspaper's real estate listings and see how eagerly hardwood floors are announced as a special bonus. This image of luxury is well deserved. A handsomely finished oak floor lends an air of dignity to a room, a feeling of warmth, and indisputable character. It is destructive to paint or carpet hardwood floors; repair them as necessary, and refinish them to reveal the wood's natural grain. If a softer surface is desired, use an area rug on top of the floor, to complement it instead of hiding it.

Most hardwood floors are made of oak strips, the most common size being 2 inches wide by $\frac{5}{16}$ inch thick. Subtle variations in the wood grain and tone contribute to the appeal of oak's natural texture. The way the sides of the strips are milled determines how the pieces should

Types of strip flooring

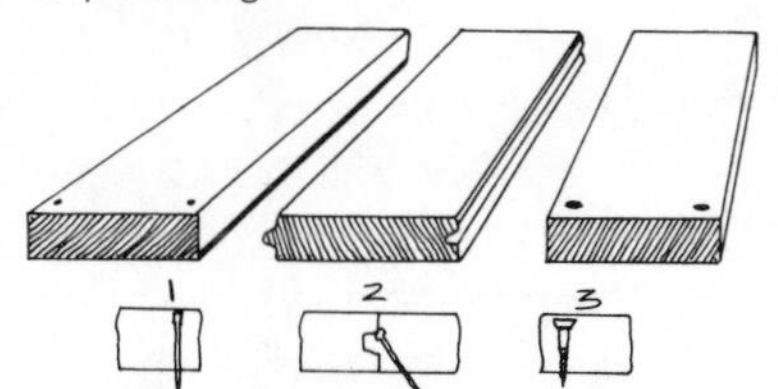

be fitted together and attached to the subfloor. Strips with a square edge should butt up flush with one another and be secured to the subfloor with facenails or screws. (The pattern of the nailheads or screw plugs becomes part of the design of the floor.) Each tongue-and-groove strip is inserted into the next strip and fastened to the subfloor with a blind nail through the tongue.

Repair. Hardwood floors are remarkably durable but not invincible. Damaged boards have to be repaired or replaced prior to the refinishing of the surface; always correct the *cause* of the damage before making the improvement. Here is how to repair the following problems:

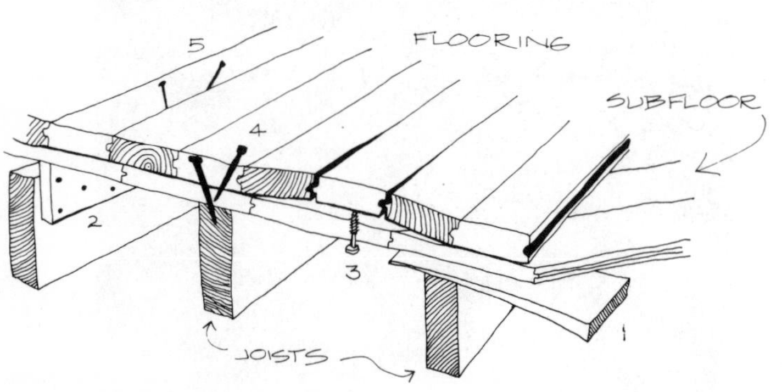

FLOORING
SUBFLOOR
JOISTS
5
4
2
3
1
SUBFLOOR
1) IF JUST ONE OR TWO BOARDS ARE LOOSE, WEDGE A SHINGLE BETWEEN THEM AND THE JOIST.
2) IF SEVERAL BOARDS ARE LOOSE, BRACE A 1×4 AGAINST THE SUBFLOOR AND NAIL TO THE JOIST.
FINISH FLOOR
3) LOOSE FINISH FLOORING CAN BE TIGHTENED THROUGH THE SUBFLOOR WITH WOODSCREWS.
4) IF THERE IS A JOIST UNDER A SQUEAKY FLOORBOARD, ANGLE RIBBED FLOORING NAILS INTO JOIST. DRILL PILOT HOLES FIRST.
5) BETWEEN JOISTS, DRIVE 6d FINISH NAILS AT SLIGHTLY OPPOSING ANGLES INTO THE FLOORBOARD CRACKS CENTERED OVER THE SQUEAK. SPACE NAILS ABOUT 6" APART.

Remedies for loose floorboards

1. *Loose boards.* Knock, nail, or screw the boards back into place if they are not too badly buckled or warped. Squeaks let you know where the trouble spot is.

2. *Water stains.* These stains darken hardwood. First correct or eliminate the source of the moisture, such as a leaky radiator or a dripping potted plant. Then bleach the stains with a solution of oxalic acid and warm water. (Oxalic acid crystals are sold at drug stores.) Brush the solution on the stain and let it soak into the wood. When the wood dries, vacuum up the crystals and sand the surface by hand or with an orbital sander. If you plan to sand the entire floor anyway, leave the bleaching crystals until you see how much of a stain remains after the major sanding operation.

3. *Protruding nails.* Tap them back into the board with a nail set. If you let nails protrude, they will chew up the belt on a drum sander or be a toe-stubbing hazard.

4. *Cracks between floorboards.* The best way to prevent cracks between first-floor boards is to install an adequate vent in the crawl space (refer to Chapter 3). Cracks occur only occasionally in West Coast homes as compared to houses in the other parts of the United States. Extreme temperature ranges combined with high humidity cause floorboards to shrink dramatically, resulting in gaps and drafty spaces between the boards. As moisture content increases, wood expands, but if floorboards abut so tightly that there is no room for expansion, the wood compresses and shrinks. When dry again, the boards are reduced to less than their original width because of the compressive stress. Fill in the cracks with wood putty, sawdust and glue, or wood splines. However, these materials are rigid and do not adapt to the swelling and shrinking of wood, so they may be only a temporary solution. Felt weatherstripping, a more flexible material that responds to changes in the width of the wood, can be forced neatly into cracks with a broad-blade knife.

Replacement. If the floor is severely warped or buckled, has bad nicks or deep scars, has urine stains, or has a hole remaining from where an outmoded floor furnace was removed, or if the border design is interrupted or has missing pieces, you have to replace the boards.

To remove tongue-and-groove floor boards, start at one end of the floor. Because of the interlocking characteristic of the planks, you have to remove the pieces in sequence. This can be a bother if the boards to be replaced are in the center of the room. Pry off the baseboard, keeping it intact, and with a crowbar gently urge up each successive board until you reach the "diseased" ones. Remove them as you did the others.

You can often find replacement boards in obscure corners of your house. Check closet floors or the portion of an upstairs room where the wall intersects a peaked ceiling and the floor extends onto the eaves. These inconspicuous areas make willing donors. Occasionally you can find hardwood flooring at salvage yards, but the most convenient source is a supplier who sells new boards (which will cost about 20¢ a running foot).

Refinishing and Sanding. Only *after* the floor has been thoroughly repaired and all the work on the walls and ceiling of a room has been completed, is it time to refinish the floor surface; otherwise, loose plaster, spattered paint, or dropped tools can ruin the floor. During the restoration process, protect the hardwood floor with an old carpet or carpet pad and drop cloth to avert unnecessary additional damage.

Floor refinishing is really not terribly difficult: you do not need the help of a professional. A dull finish that does not require a complete overhaul can be rejuvenated with mineral spirits or turpentine. Rub out rough spots with fine steel wool; use diapers or paper towels to distribute and absorb the liquid. Complete refinishing involves smoothly sanding down the uppermost layer of the wood, applying a stain if a change of color is appropriate, and applying a protective coat of oil or plastic to arm the wood against wear.

If you removed linoleum or a glued-down carpet to get to the hardwood, you have to clean off the mastic and paste before you can begin sanding. This involves using chemical solvents, scrapers, and industrial sanders. To prepare for sanding, remove all furniture, draperies, and other surfaces that can trap fine dust. This prevents the dust from drifting back on to the floor while the final sealer is drying and ruining the smooth surface. Close doors to adjoining rooms and halls to contain the dust, and seal the doors with 2-inch-wide masking tape. Keep windows open for adequate ventilation (here is where double-hung windows really pay off!).

Use power sanders to remove the old finish and smooth the surface. A drum sander is recommended for the center of the room, an edger for around the perimeter, where the floor and walls meet. You can rent these machines at many hardware stores.

Sand across the floor in a forward and backward direction in three passes. Sand only *in the same direction* as the lay of the boards and the grain of the wood or you will thoroughly ruin a valuable floor. Even one scrape of the sander against the grain will leave unsightly gouges in the wood. Try to prevent this. Sometimes it takes a lot of muscle to keep the sander on a straight course, especially if you are using a heavy-duty drum-type industrial machine. Make the effort though, because the result is worth it.

To properly sand, on the first pass use a coarse sandpaper (grit size 2½ -1½) to break up the old finish. For the second pass, use a medium sandpaper (grit size 1-0) to remove scratch marks left by the coarse paper. Finally, on the third pass, use a fine sandpaper (grit size 2/0-5/0) to remove scratches left by the medium paper and leave the floor perfectly smooth.

As you sand, remember the following points:

1. Oak strips that were once 5⁄16 inch thick do not allow much leeway for sanding, especially if they were sanded before. (You can identify 5⁄16-inch board by facenails 7 inches apart.) And tongue-and-groove floorboards are ruined if they have been sanded down to the tongue. If in doubt, remove a board and see how much floor you have left. Also note how much allowance there is between the top of the board and the nailhead.

2. The fine dust that accumulates in the sanders' collector bags is highly combustible, and sparks are likely to be sucked up into a bag during the sanding operation. Empty all collector bags into boxes or cans and place them outside, where they can do no harm should a smoldering spark ignite.

3. Drum sanders draw a lot of current, so make sure that the voltage required by the rented machine matches your house's output. Also, you may need a heavy-duty extension cord and a three-prong adapter.

4. As stated, a drum sander is a powerful machine and seems to have a mind of its own, and it involves a great deal of strength to operate. Never let go of the sander while it is running, or it might try to run away across the grain of your floor. And do not let it stand in one spot while it is running or the sander will grind a hefty crater in the wood.

5. Cleanup after sanding is critical to the success of the finish. Vacuum the floor, baseboards, windowsills, bookshelves, and everything in the room that could possibly sequester dust.

Finishing. After sanding, the color of the floor can be left as is or darkened with a stain. Remember that with just a clear finish rather than a stain, the floor will look darker than it did as raw wood. Prove this by spreading mineral spirits (benzine) or turpentine over several square feet. The wood will have about the same tone when wet as it will if finished with a clear coating.

If you decide to use stain, remember that the stain will look darker on a small sample than it will on the floor. An overly colorful commercial stain can be diluted with turpentine for a more subtle effect, but keep a record of the proportions used, in case you run out between coats. Be sure that the stain you select is chemically compatible with the finish you have in mind. *Never* use a varnish stain because it obscures the grain like a sheet of formica.

Decide carefully which finish to use. There are two basic types: surface film (shellac, varnish, quick-dry varnish, polyurethane), which coats the top of the floor and withstands food and traffic; and penetrating (penetrating sealer, oil), which seeps into the wood, filling the spaces between the uppermost fibers. A penetrating finish

imparts the soft luster associated with old wood floors, but it does not provide the abrasion resistance of a surface film finish.

The best wood floor finish for a household is polyurethane because it resists abrasion, is easy to maintain, and comes in gloss or satin sheen. The satin finish shows less dirt and is easier to maintain, and it gives you the option of later switching to a gloss finish if you so desire. Although it is one of the most expensive finishes (about $20 a gallon), it is far more durable and longer lasting than many others. For best results, apply the polyurethane in several coats rather than a single thick one. Polyurethane is incompatible with certain stains.

SOFTWOOD

Softwood floors, found in kitchens, pantries, bedrooms, and second stories of older homes, are typically made of tongue-and-groove Douglas fir strips; although pine boards and square-edge joints were used too. Unlike oak boards, softwood boards are as broad as 4 inches wide and go right up to the wall, without a border.

Normally fir is associated with rough construction and is not considered a finishing material. However, fir does have special qualities. First, fir is one of the harder softwoods, really wearing a lot better than others. It is also a lot less difficult and less expensive to replace damaged fir portions than oak. Visually, its pronounced, open grain and reddish nuances offer a rich display. It is far superior to pine because pine is exceedingly soft, has a muted yellowish-white appearance, and does not present any prominence in its grain pattern. Also, pine does not refinish as readily and smoothly as fir.

You repair, replace, sand, stain, and finish a fir floor as you do an oak floor. But remember one critical point: because fir is softer than oak, be extra careful during the sanding process. Be more delicate in your touch; a heavy drum sander would love to eat a gaping hole in the wood if given half a chance. Start with a medium-grade sandpaper, unless the floor has a heavy coat of gummy finish on it. For best results, make the final finish a polyurethane coating, to add an extra shield of hard protection.

CERAMIC TILE

Ceramic tile is the original flooring for practically all bathrooms. The bathroom itself was introduced during the Victorian era, as a separate room for bathing instead of just a zinc tub in front of the fireplace. At first the bathroom floors were softwood, splatter painted in five different colors; but with the fashionable concern for hygiene at the turn of the century, tiles (resistant to moisture seepage and bacteria collection) were quickly incorporated into bathroom floors and walls. Pastel color tiles were not common until the 1920s.

Impervious (nonporous) moisture-repellent bathroom tiles were the grand introduction around 1910. The design most in vogue at this time was the small, white hexagonal tile, resembling a pattern of chicken wire. (These tiles were also used on kitchen counters.) Often, as added drama to the repetitious pattern, black hexagonal tiles were placed at regular intervals throughout the pattern of the floor. This was called the spiral-weaver pattern. An unfortunate feature of this and any tile floor reliant upon white grouting is the grout's natural tendency to absorb moisture and dirt and darken with age. This is not a displeasing feature in itself, except that the grout does not darken uniformly throughout the floor's coverage. Areas around the edge of the walls or under the fixtures do not collect water and dirt as do the grouted areas in the center of the walked-upon floor, even though it may be quite clean. The best solution is to dye the grout black so its appearance is uniform throughout the floor. (Tile suppliers sell the dye.)

Sometimes regrouting and replacement of tiles are necessary. This is a simple job and requires no special skills, tool, or patience. Many tile suppliers still carry the white hexagonal tiles because they were popular through the 1930s. To replace tiles, be sure that the floor area is clean, smooth, and dry. Then follow the manufacturer's or supplier's grouting instructions.

Quarry tile, the terra cotta (brick-colored) square or ornately-shaped tile, is typical of Mediterranean-style homes. This smooth, semiporous tile is usually much larger than the 2-inch hexagonal tiles. Quarry tile was often used as both an interior flooring for entry ways and some exterior surfaces. Always replace quarry tiles with quarry tiles; fortunately, quarry tile is still popular and widely available today. Replace and regrout as you would for other types of tiles.

RESILIENT FLOORING

Resilient flooring is made of a pliable material like vinyl, asphalt, linoleum, rubber, cork, and various synthetics and is available in tile or sheet form. It is usually used in the kitchen or bathroom. It is difficult to recommend specific guidelines for replacing worn-out linoleum or switching to resilient flooring because no historical rules really apply, since the materials are all modern. Consider the ultimate end product. A floor sets the stage for the presentation of such detail as wall coverings, furnishings, and fixtures. In no way should the floor attempt to fight the other elements in the room; resilient cover should be subtle in its appearance.

The wisest decision is to use a nonobtrusive color, pattern, or texture. A solid neutral floor without any frills will direct attention to the more authentic and prominent architectural features of the room. Patterns make small rooms look larger, and in many cases a patterned floor can give a room an awkward and unsettled feeling. *Never* use a pattern that attempts to imitate a material like brick, marble, wood, and so forth. It will look blatantly phony and detract from the room's other elements.

Resilient tiles are somewhat easier for an amateur to install than sheet vinyl, but the sheets offer better hygiene because they have no dirt-collecting cracks or gaps. And with sheet vinyl the edges can be curved and coved up the edge of a wall to serve as a baseboard. This "baseboard" corners dirt and moisture and controls mop sloshing. However, coved bases are highly inappropriate when they cover up shoe molding or if they are so high it looks as if the floor is climbing up the wall. Install the sheet vinyl flat so it abuts the wall in the same manner as tile would. For a finished edge where the resilient flooring meets the wall, use an ample strip of *wood* molding, not the thin metal strip used with modern floor installation.

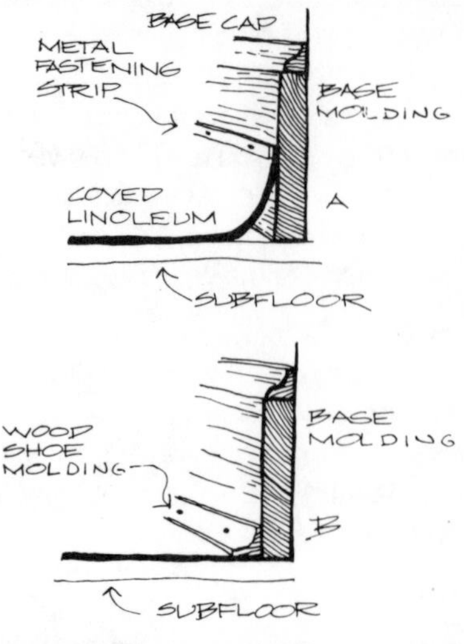

Method of fixing flexible flooring

The metal strip looks shiny and skimpy in a location where something with a subdued finish and more girth is appropriate; and it is not in keeping with the character of the rest of the room.

The most difficult restoration problem with resilient floor covering is removing it to refinish a wood floor. As we mentioned eariler in the chapter, use chemical solvents, scrapers, and industrial sanders. This is sometimes frustrating and trying because years of pressure have settled the gummy mastic firmly in place.

(CHAPTER 11)

INTERIOR WALLS AND WOODWORK

The quality of a wall's surface, the quantity and quality of interior trim, and interior ornamentation all determine the character of a room. If the restored wall and ornamentation treatment is consistent with the original design of the house, the room will provide both interest and visual comfort. On the other hand, if the original treatment of the room is disregarded, the room will be cold and boring.

(PLASTER)

Plaster is the universal wall surface for turn-of-the-century houses. It provides a plain but not bland background to architectural details like moldings and beams. Plaster frequently needs repair because it is a rigid material tied into a flexible wood framework. As the house settles, the wood members shift, including the lath strips nailed to the studs, but the plaster attached to the lath does not. It cracks. Plaster is also prone to brown spots from leaking water, chips from picture fasteners that pull away, and holes from plumbing and electrical work.

Minor cracks and chips do not justify the expense of switching to another wall surface, because any amateur can easily fix them. Minor plaster repair is similar to that for stucco. As with stucco, repair begins with correcting the cause of the problem: leaks, broken or rotted boards, and so on. If you do not correct the cause, your surface repair will be only temporary.

SHEETROCK

Very large holes and extensive deterioration of a plaster wall call for complete replastering, which should be done

Removing wood ornamentation

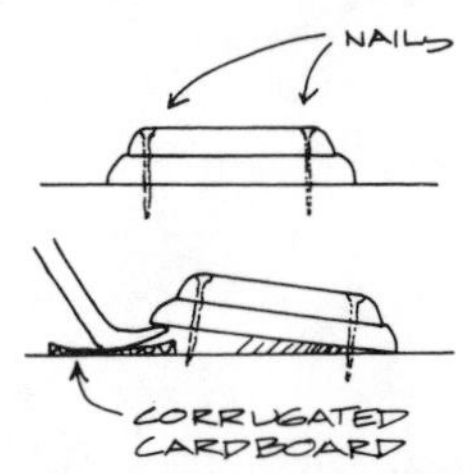

WHEN USING A PRYING TOOL, WORK SLOWLY AROUND PERIMETER OF ORNAMENT IN STAGES. APPLY PRESSURE BENEATH NAILS TO PREVENT SPLITTING. SLIP THIN WOOD OR CARDBOARD BENEATH PRYBAR TO CUSHION THE STRUCTURE.

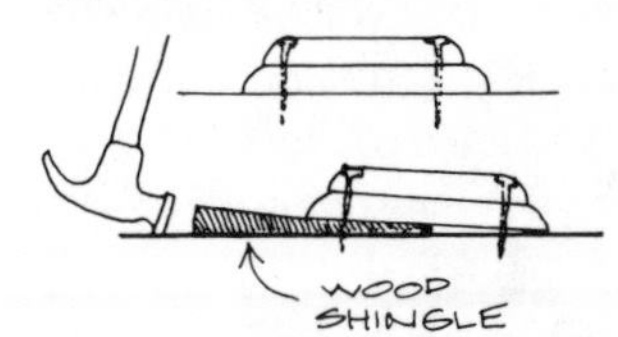

AFTER OPENING A SMALL CRACK WITH A PUTTY KNIFE, DRIVE ONE OR MORE PIECES OF WOOD SHINGLE UNDER ORNAMENT, FORCING IT UP. COMPLETE REMOVAL WITH CLAWHAMMER, OR PRYBAR IF NECESSARY.

by a professional. A more modern solution to major plaster problems is sheetrock, also called gypsum board, which is like a solid sheet of plaster and comes in 4 × 8-foot panels of varying thicknesses. The most commonly used residential board thickness is ½ inch, although in some areas and for multiunit dwellings, the law often requires ⅝-inch thickness. The installation of

A) HAIRLINE CRACKS CAN BE FILLED WITH A THIN MIXTURE OF PATCHING PLASTER OR WALLBOARD JOINT CEMENT. RUB THE CRACK LIGHTLY WITH VERY COARSE SANDPAPER, THEN BRUSH OUT LOOSE MATERIAL. FORCE THE FILLER DEEP INTO CRACK WITH FINGERTIPS, THEN SMOOTH SURFACE WITH V STROKES OF A PUTTY KNIFE.

B) SMALL CRACKS SHOULD BE OPENED AND UNDERCUT WITH A SPECIAL TOOL OR A CAN OPENER. THIS PREVENTS THE CRACK FROM REAPPEARING AFTER FILLING & PAINTING.

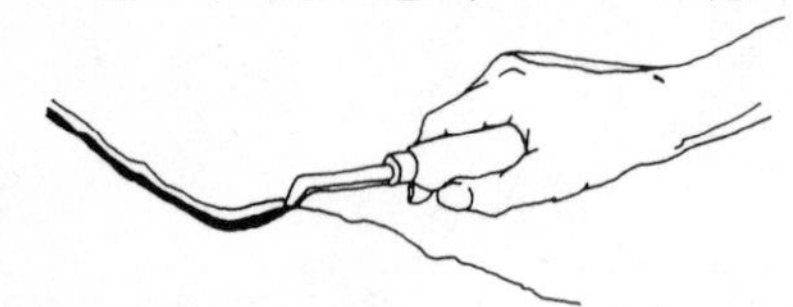

CLEAN OUT LOOSE MATERIAL FROM CRACK. FORCE FILLER DEEP INTO CRACK FROM 2 DIRECTIONS, THEN LEVEL SURFACE WITH V STROKES.

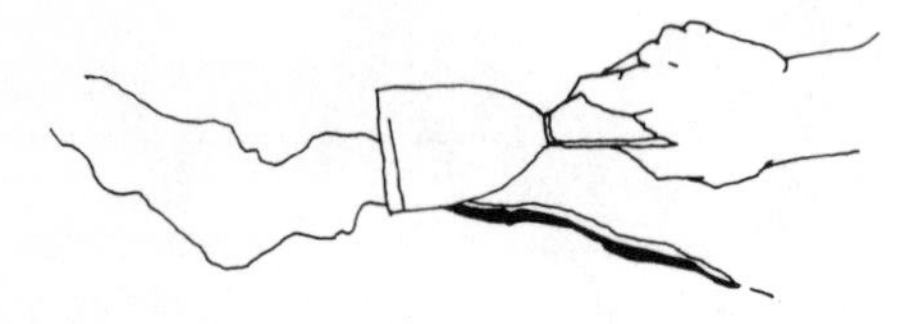

C) WIDE CRACKS AND HOLES LARGER THAN A FEW INCHES IN DIAMETER SHOULD BE FILLED WITH 3 LAYERS OF PATCHING PLASTER. UNDERCUT THE EDGES OF THE HOLE OR CRACK BACK TO THE LATH, MAKING SURE THAT THE LATH IS FIRMLY ANCHORED TO THE STUDS. (IF NOT, OR IF THERE IS WATER DAMAGE, SEE ILLUSTRATION ON EXTENSIVE REPAIR.)

DAMPEN CRACK WITH SPONGE. ADD FILLER TO HALF THE DEPTH OF THE CRACK, SCORE THE FILLER COAT WITH COMB OR PUTTY KNIFE, THEN ALLOW TO DRY. BRING SECOND COAT TO WITHIN 1/8" OF THE WALL SURFACE. LET DRY, APPLY FINAL COAT WITH TROWEL OR WIDE-BLADED PUTTY KNIFE.

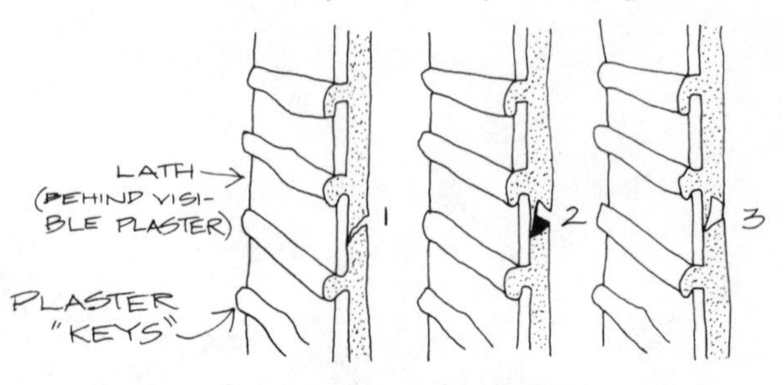

1) CRACK IN PLASTER
2) CRACK OPENED & UNDERCUT
3) FILLING COMPLETE

sheetrock, unlike plastering, can be handled by the average handyperson.

Use sheetrock to fill in large areas where plaster is dilapidated; do *not* use it like paneling, projecting out from the wall and covering up or eliminating architectural details.

Guidelines.

1. Carefully remove any molding, cornice work, or trim from the wall area where plaster will be replaced by sheetrock, and store it safely. You may be able to leave the shoe molding in place if the plaster can be cut below the top of the molding, with the sheetrock installed behind it.

2. Pull off the plaster and its lathing back to the studs. As with many demolition jobs involved in restoration, it does not take much skill to rip something out, just time and aggression. This is a dusty operation, so remove all furniture and draperies from the room, close the doors and seal the cracks with 2-inch-wide masking tape, and open the windows. Use an inexpensive fiber mask, or make one out of a scarf, towel, or half a brassiere. At the end of the work day, sweep and vacuum up the dust that settles.

3. Install the sheetrock so it fits into the space once filled by the plaster. If possible, match the sheetrock thickness to that of the plaster wall. The sheetrock should not create a bulge or too deep a depression in the wall. Remember, think of sheetrock as solid plaster, not as paneling.

4. If the surface texture of the sheetrock is not the same

Extensive plaster repair

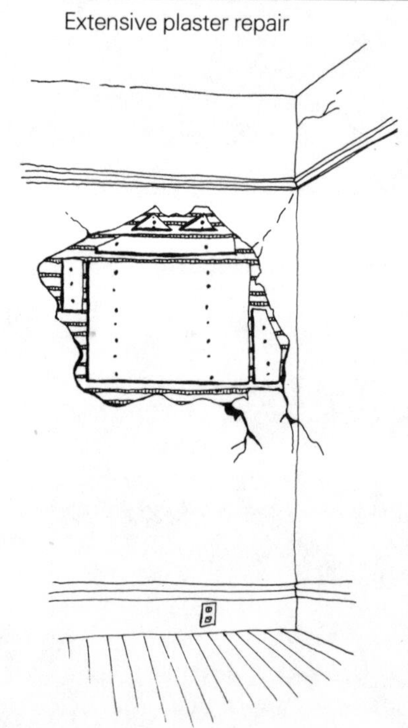

WHEN AN EXTENSIVE SECTION OF PLASTER IS LOOSE OR CRUMBLY, PULL THE LOOSE MATERIAL OFF THE WALL, SAVING LATH IF POSSIBLE. NAIL SHEETROCK TO STUDS, FILLING IRREGULAR GAPS WITH SMALL SCRAPS NAILED TO LATH. SHEETROCK SHOULD BE SAME THICKNESS AS OLD PLASTER OR A FRACTION LESS.

APPLY TWO OR MORE LAYERS OF THICKLY MIXED PATCHING PLASTER, PRESSING FIRMLY INTO CRACKS & LATH AND STOPPING JUST SHY OF THE ORIGINAL THICKNESS. LEVEL PATCH WITH A FINAL THIN COAT OF PLASTER.

as the existing plaster, match the sheetrock with coat(s) of special spackling compounds (but first practice on a scrap). Matching the surface finish is especially important in Period Revival houses, where the plaster was heavily textured.

5. Replace the molding or trim to its original position on the wall. This is extremely important because blank sheetrock walls look rather insipid, but when combined with architectural features they are meaningful, particularly in high-ceilinged Victorians. An expanse of sheetrock on a 10- or 12-foot-high wall will look like a drive-in movie screen if the trim is not returned to its rightful place.

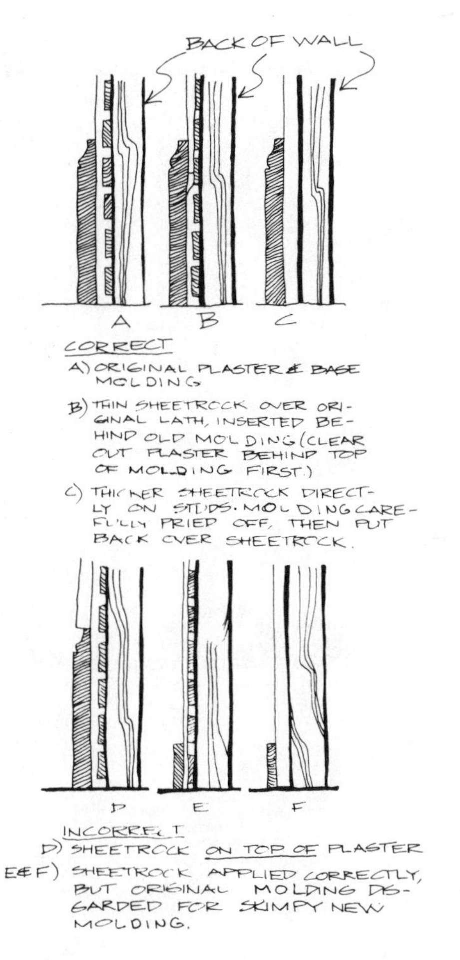

Sheetrock — Do's and Dont's

(WAINSCOTING)

Judging by the walls of modern tract homes, you would never think that wood could be used as a wall surfacing material. Yet in older homes wood walls and other interior features are quite common. In fact, they are one of the main reasons these vintage residences feel so warm and welcoming inside. Wood paneling absorbs light and provides a deep and rich subtle glow, whereas stark white nonwood surfaces used throughout reflect only coolness without significant drama or contrast.

Wainscot styles and vocabulary

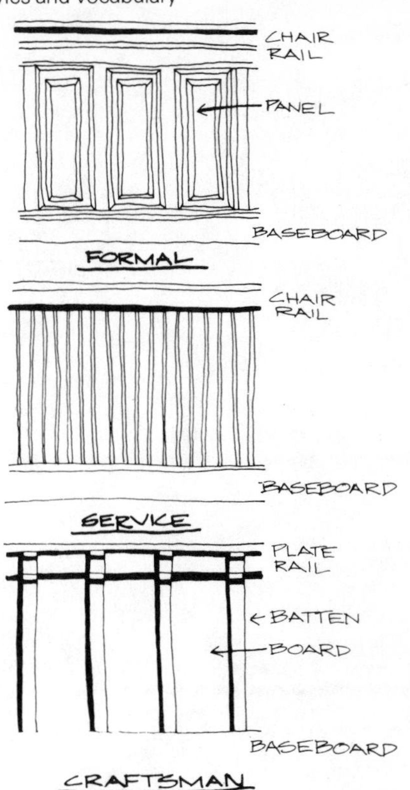

Many older homes have a mixture of wood and nonwood surfaces. A skirt of wood (wainscoting) around the perimeters of rooms like the living and dining rooms serves both practical and aesthetic purposes. These half-wall panels run about 3 to 4 feet up the wall from the floor level and act as a bumper against chairs and foot kicks. The protective wood makeup offers easy maintenance, as with any wood floor. Visually, wainscoting helps break up and somewhat define the light plaster areas. Wainscoting often differs from house to house and certainly from architectural style to style. Wainscoting is also typically complemented by such other wooden accents as beams that create *coffered ceilings* (ceilings with recessed panels) and prominent moldings of the period. These accents should also be preserved and restored, as described in the chapter's later Trim section.

RE-CREATION

If wainscoting is still in place, by all means leave it there. If the wainscoting has been removed, it can be re-created as shown. Sometimes the wainscoting is rather elaborate and somewhat difficult to reproduce because its "outdated" style is difficult to locate. Some distributors *do* carry certain historical designs, so try your best to locate these outlets. If replacing formal, paneled wainscoting will be too expensive for your budget, reconstruct it as you would wood porch handrails (see Chapter 8). With a little examination you can easily determine the molding and panel parts. Often, clever use of veneered plywood, flat moldings, and quarter-round pieces will duplicate your original molding quite nicely. (Of course, re-creating

Recreating a Wainscot

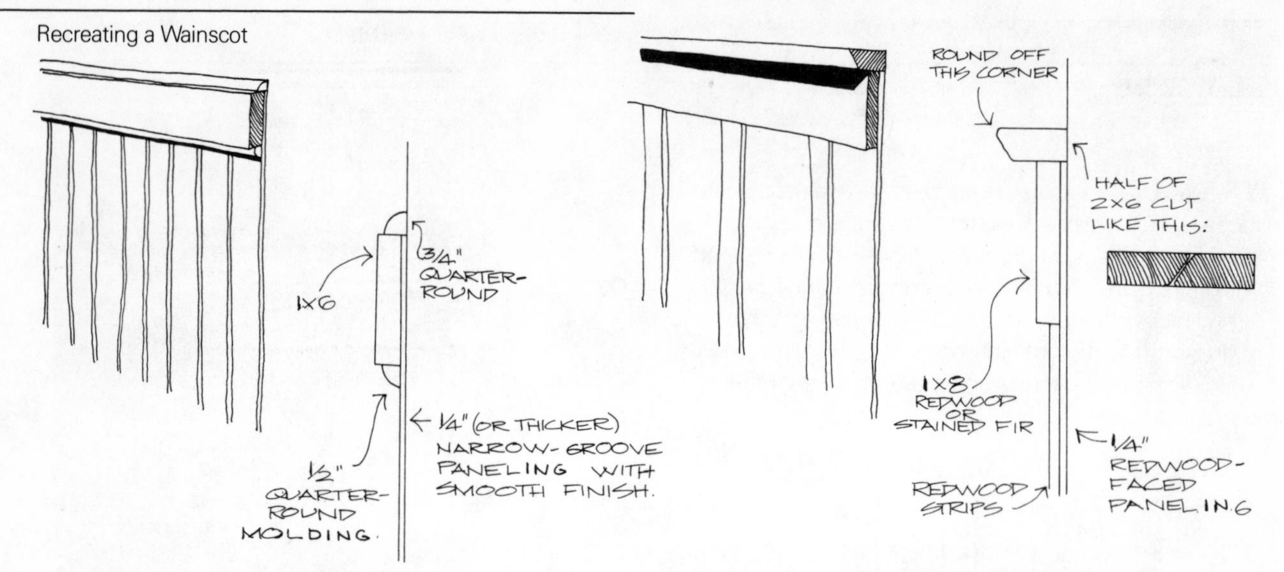

A) RE-CREATING A SERVICE STYLE WAINSCOT: CHOOSE A STRAIGHT-GRAINED WOOD PANELING WITH REGULARLY SPACED GROOVES 2"-4" APART. IF PANELING COMES IN 8' SHEETS, USE 4' SECTIONS TO SAVE WOOD. PLACE SECTIONS ABOVE BASEBOARD MOLDING & NAIL TO STUDS. BUILD UP TOP MOLDING FROM QUARTER-ROUND MOLDING & 1X6 BOARDS, AS ILLUSTRATED. FILL NAIL HOLES, THEN STAIN MOLDING TO MATCH PANELING. SEE SECTION ON PAINT STRIPPING FOR HINTS ON FINAL FINISH.

B) RE-CREATING A CRAFTSMAN STYLE WAINSCOT: NAIL HALF SHEETS (4'x4') OF SMOOTH-FINISH REDWOOD FACED PANELING ABOVE BASEBOARD, WITH GRAIN POINTING VERTICALLY. CUT 2X6 LUMBER IN HALF LENGTHWISE AT A 60° ANGLE. (YOU WILL NEED A CIRCULAR SAW FOR THIS) AND BUILD TOP MOLDING AND PLATE RAIL AS ILLUSTRATED. PLACE SMOOTH-SANDED STRIPS OF WELL-DRIED REDWOOD LATH (BUY 6" LATH & CUT INTO 3" STRIPS) AT 1- OR 2-FOOT INTERVALS OVER PANELING COVERING JOINTS BETWEEN PANELS.

simple and self-defined wainscoting, such as that illustrated, is much easier.)

FINISH

If the wainscoting is unpainted, *do* leave it that way. Chances are that this is really authentic wainscoting and meant to look as you now see it. And if the skirt is a rare remaining example of graining—a nineteenth century skill by which one wood was artfully painted to look like another—that too should be left as is. This may seem twentieth century in appearance, but the process does not belong to pseudomaterial fads of the post-World War II era; it really is an expression of late nineteenth century design interpretation.

If the wainscoting has been painted, you can repaint it, strip and finish it, or arrange to have it grained. Look in your area for craftspeople who specialize in these distinctive skills. Please, do not fall for the "quickie" antiquing tricks some people offer. This is nothing more than an obviously phony two-tone paint and glaze job. And shun the sadistic process called "distressing," which is the rape of good wood by beating it with hammers and chains to make it look aged and worn.

(OTHER WOODWORK)

REPAIR

Existing woodwork may require repair prior to refinishing.

Cross section of a plywood panel

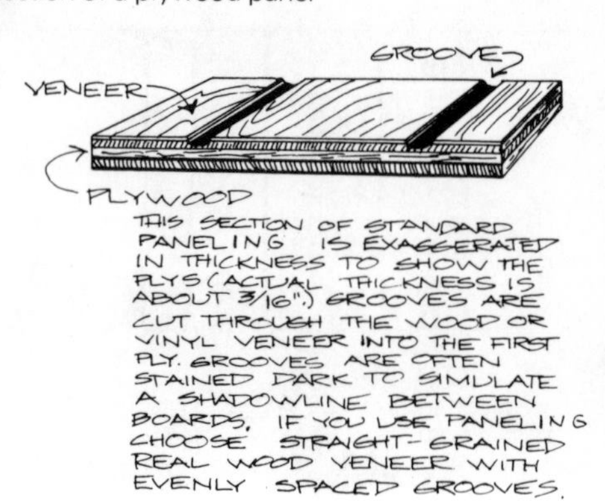

GROOVE
VENEER
PLYWOOD
THIS SECTION OF STANDARD PANELING IS EXAGGERATED IN THICKNESS TO SHOW THE PLYS (ACTUAL THICKNESS IS ABOUT 3/16".) GROOVES ARE CUT THROUGH THE WOOD OR VINYL VENEER INTO THE FIRST PLY. GROOVES ARE OFTEN STAINED DARK TO SIMULATE A SHADOWLINE BETWEEN BOARDS. IF YOU USE PANELING CHOOSE STRAIGHT-GRAINED REAL WOOD VENEER WITH EVENLY SPACED GROOVES.

To replace damaged pieces, bring a sample to a lumber yard that stocks a variety of interior-finish lumber to match type, grain, size, and stain. Redwood, fir, and mahogany are fairly easy to find, but gum wood, such as eucalyptus (so typical of Prairie School Revival), is no longer available. Birch is really about the best substitute. The grain of birch approximates that of gum and is light enough in color to take matching staining. If the original pieces are thicker than standard replacement stock, you may have to cut new pieces to size. A less costly solution in nonprominent areas is to buy a thin piece and back it with the proper thickness of wood so it matches the original setting. Plywood is perfectly sufficient as the

backing and is economical when refurbishing areas that are not closely scrutinized.

STRIPPING AND REFINISHING

Refinishing woodwork can make a dramatic improvement in the character of a room. However, stripping is a tedious commitment that should be undertaken only with selectivity and forethought. It is advisable to strip, paint, or varnish woodwork when:

1. The architectural style calls for natural wood.

2. There is fine hardwood underneath, and you have the time and patience to do the job right. (Delinquent specks of paint or color negate the improvement.) As with floors, hardwood was usually reserved for the formal rooms, and softwood for the family room and service areas.

3. Stripping would remove paint that has obscured detail, carving, or molding.

4. There is softwood underneath but the paint is alligatored, and stripping is more expedient than scraping.

5. There is a varnish or shellac finish that has darkened unattractively with age, so a fresh, clear coat is desired.

Stripping and refinishing interior woodwork involves techniques similar to those used for refinishing furniture, but woodwork has the advantage of being a relatively flat, stationary surface that can be worked on without additional bracing. However, stripping is unavoidably messy and clouded by unpleasant fumes or accompanied by a piercing noise. Familiarize yourself with the options described below. Choose your weapon, save up old newspapers, and be sure to ventilate the room.

Chemical Removers. Chemical removers are applied with a brush and left in place long enough to buckle the layers of unwanted paint so they can be scraped off in a continuous ribbon with a wide-blade putty knife. There are at least seven different types of chemical compounds. The most appropriate remover for interior woodwork is nonflammable and heavy bodied and can be rinsed off with water. It is thick enough to hold on vertical or irregular surfaces and contains methylene chloride, to minimize fire hazard. Typical brands are Zip-Strip and Jasco, available for about $12 a gallon.

The two rules for successful chemical stripping are: (1) do not be stingy with materials, and (2) let the remover sit long enough to cut through all the paint.

Although most home-repair books recommend stripping only a small area at a time, some professionals do large sweeps (as much as 50 square feet) with great efficiency. This is advisable for the amateur once the knack has been acquired. If any of the large soaked areas dry out before you get there with a blade, wet them down again with more remover.

Heat Removers. Heat removers are appliances that literally melt the paint away. Several kinds are available. You can buy a semienclosed electric coil, similar to the heating element on top of a stove, at a paint store, or you can rent one from a tool-rental shop. Do *not* use propane blow torches for the paint-removal process; not only do they easily scorch the wood underneath, they are a risky fire hazard.

Another option combines heat and chemicals. Use a wallpaper steamer to accelerate the effect of a water-rinsable paint remover. Apply the stripping compound, preferably one with a high methylene chloride content, and let it sit for 15 minutes. Apply the steam through the pan of the steamer, moving it slowly across the wall, about 1 square foot a minute. Follow with a wide-blade scraper; you can remove four or five coats promptly. Do not use steam with chemical removers that contain carbon tetrachloride or benzene.

Manual Stripping. The manual scraping of woodwork involves more physical effort than heat or chemical methods. But health constraints, poor ventilation, or personal preference may preclude the heat and chemical removers, because nothing short of a gas mask can prevent some intake of the noxious fumes they emit. Also, successive layers of old, rock-hard enamel may not yield easily to chemicals. Manual scraping can, however, cut through paint layers and into the pristine wood laying below the paint-impregnated outer wood fibers.

A hook-type scraper is the basic tool. Buy a 2½-inch-wide bladed scraper for wide surfaces and fast cutting, and a scraper that is small enough to fit into tighter spots. Follow this technique:

1. Use two hands when you can. Use one hand to pull on the end of the handle, so you can control the speed and cutting angle; use the other hand to push on the blade end, controlling the pressure. Determine the most efficient blade angle, and use only as much force as necessary, to conserve your energy.

2. Sharpen blades often for easier, smoother scraping. Replacement blades are available, but most blades can be sharpened with a fine-tooth flat file. When using the file, maintain the original bevel of the blade.

3. Scrape with the grain, especially on woods with pronounced grain, like fir. Scraping across the grain or on the end grain will splinter and gouge the wood.

4. If you dislike sounds similar to fingernails scraping on a chalkboard, wear earplugs to lessen the impact of chilling, high-pitched vibrations the scraper makes. Wear goggles to protect your eyes from flying paint specks.

Stain/Finish. Once the wood has been stripped, you must apply a stain or a finish to protect the wood. Stain selection is a matter of architectural style and interior decor. The dark stains, appropriate to Colonial Revival houses, are dramatic but sometimes appear rather somber if prevalent through the room. A slight divergence from accuracy may be warranted for modern taste as well as for permitting more incoming light to be reflected more warmly.

Finishes are film forming (varnish, lacquer, and shellac) or oil (boiled linseed, Danish, and tung oil). The secret to using film finishes is to thin them first. Three thin coats dry faster and harder than thick coats, and a thin first coat seals the wood properly.

Thin varnish with turpentine (four parts varnish to one part turpentine), lacquer with lacquer thinner, and shellac with a denatured alcohol. An old nylon stocking is a good applicator that leaves no bristle marks. Oils are typically applied with steel wool or soft cotton, but for large surfaces you can brush them on if each of the three coats is very thin. Whichever finish you select, always work with the grain of the wood, and always allow plenty of drying time—usually 24 hours—between each application. Adequate ventilation is crucial when using these volatile fluids.

[LINCRUSTA-WALTON]

Lincrusta-Walton was a heavy, embossed, cardboard wall covering, imported from England and Belgium in the late nineteenth century as wainscoting and wall covering for Victorian and Colonial Revival parlors and hallways. It came in large rolls and was soaked in water for several hours before installation. Unfinished Lincrusta-Walton

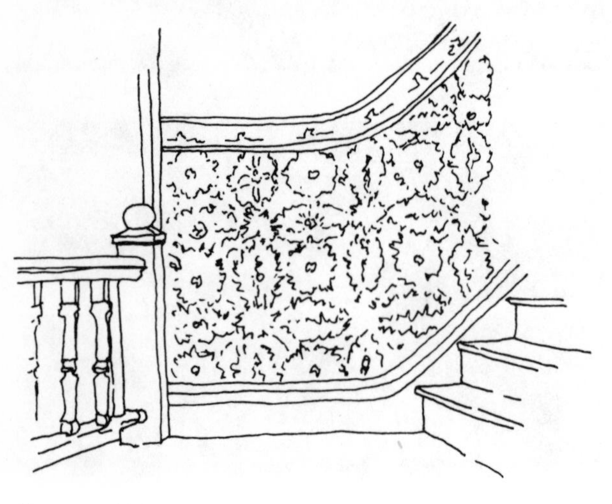

Lincrusta-Walton

was beige, but in homes it was usually coated with a glossy, brown varnish. Subsequently, as Victorian tastes changed, the golden surface was often buried under layers of paint.

A house that has Lincrusta-Walton boasts a real antique. No longer commercially available, the embossed cardboard is extremely valuable and should be retained and shown off. You can mend patches deteriorated by water by making a plastic mold from an unharmed portion and filling the mold with papier-mache. (This ingenious technique was developed by Agnes Pritchard, plasterer, in association with San Francisco Victoriana.)

[PANELING]

Paneling is a modular wall surface that comes in an untold number of designs in an astounding range of natural and special effects. Paneling falls into three basic categories: (1) solid wood, (2) plywood with veneer, and (3) nonwood, like polystyrene, plastic laminates, fiberglas, hardboard, and cork. Modern paneling is *not* at all similar to the original wall surfaces of any of the architectural styles discussed in this book, although a nonmanufactured version of solid wood wall panels is found in some Victorians, Colonial Revivals, and First Bay Tradition houses.

Paneling is a gruesome mistake when it is used to cover up original architectural features that are still substantially intact. However, if a wall in a less important room has been devastated by earlier misguided remodeling efforts, paneling can be an acceptable solution if you use the following guidelines:

1. Always use real wood products. Imitations look cheap, even if they cost dearly.
2. Select paneling with regular spacing between the vertical grooves. So-called random spacing looks contrived rather than interesting.
3. Select a wood with color and grain sympathetic to the original architecture, as indicated by other features in the room.
4. Avoid flamboyant grains, like birds-eye maple or pecky cedar, that compete rather than work with the original features in the room.

Paneling can also be used to re-create the proportions of long-lost wainscoting, as illustrated previously.

REMOVAL

You will probably want to remove unwanted paneling, to capture the original character of a room. Paneling is either nailed to the studs or glued in place with a contact cement. Some panels are additionally attached to each other with tongue-and-groove joints. Look for nails hidden in the tongue or recessed in the V groove.

To remove paneling you will need a pry bar, a hacksaw blade, and brute strength. Work the pry bar under the edges of each panel to loosen it. Cut nails you can reach with the hacksaw blade. Be prepared to completely renovate the wall because damage to the original plaster is almost certain. The most practical replacement is to install sheetrock against the studs and reinstate the trim as needed.

[PAINT]

Paint is properly applied to plaster and sheetrock, and in some cases wood. The advent of latex paint has markedly simplified the painting process. Latex, a water-based, as opposed to oil-based paint, is thinned with

Anatomy of a room

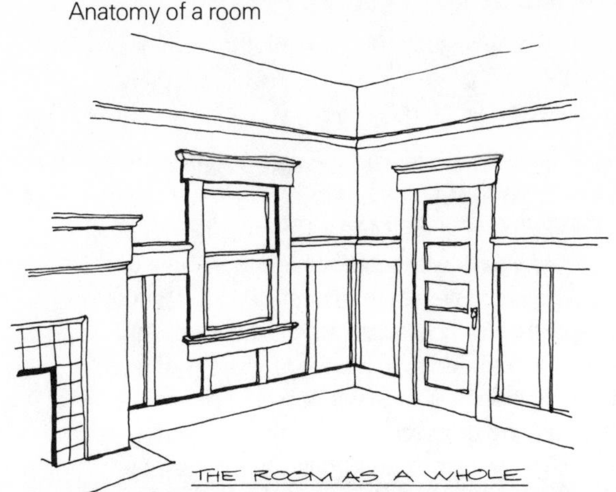

THE ROOM AS A WHOLE IS THE FOREMOST CONSIDERATION. THE COMPONENTS THAT MAKE UP A ROOM SHOULDN'T DEMAND TOO MUCH INDIVIDUAL ATTENTION. COLOR CHOICES SHOULD CREATE HARMONY. IF YOU FEEL A NEED FOR BRIGHT COLORS, FILL IT WITH ART-WORK & FURNITURE. ONE WAY TO CRE-ATE HARMONY IS WITH A ONE-COLOR PAINT SCHEME, BUT CHOOSE A LIGHT COLOR, UNLESS YOU LIKE IT GLOOMY.

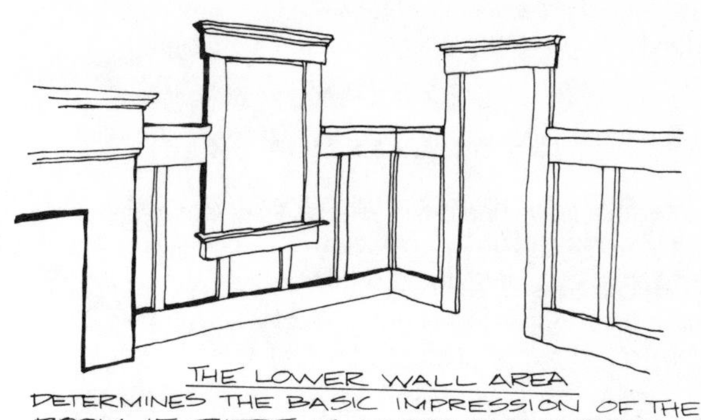

THE LOWER WALL AREA DETERMINES THE BASIC IMPRESSION OF THE ROOM. IF THERE IS UNPAINTED WOODWORK, LEAVE IT THAT WAY.
IF YOU DESIRE A 2-COLOR PAINT SCHEME, THE UPPER & LOWER WALL SURFACES SHOULD RECEIVE THE 2 COLORS. IF THERE IS NO MOLDING DIVIDING UPPER FROM LOWER, IT CAN BE ADDED. TO UNIFY THE ROOM, THINK OF DOOR & WINDOW TRIM AS PART OF THE LOWER WALL, & PAINT IT THE SAME COLOR. THE LOWER WALL AREA SHOULD BE DARKER THAN THE UPPER WALL TO VISUALLY SUPPORT IT.

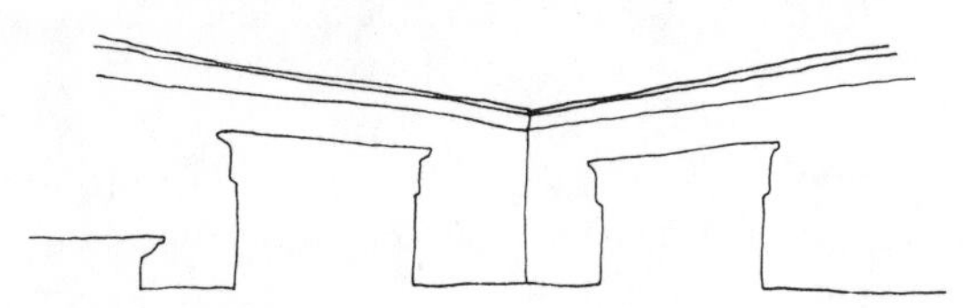

UPPER WALL AREA— THE CORNICE MOLDING SIGNIFIES THE END OF THE WALL & THE BEGINNING OF THE CEILING, EVEN IF THERE IS MORE VERTICAL WALL SURFACE ABOVE IT. IN OLDER HOMES WITH HIGH CEILINGS, THE POSITION OF THE CORNICE MOLDING KEEPS THE EXPANSE OF WALL SURFACE AT A MORE HUMAN SCALE. ACKNOWLEDGE THIS FUNCTION WHEN SELECTING PAINT RE-GARDING THE CORNICE MOLDING AS PART OF THE UPPER WALL AREA AND USING THE SAME COLOR FOR BOTH.
LEAVE THE CORNICE MOLDING UNPAINTED ONLY IF THE OTHER WOODWORK IS UN-PAINTED, OR THE CEILING & UPPER WALL ARE DIFFERENT COLORS. OTHERWISE, THE MOLDING WILL LOOK LIKE A BOLD STRIPE INSTEAD OF THE JUNCTION BE-TWEEN CEILING & WALL.

ACCENTS ARE OPTIONAL. WHEN USED, THEY SHOULD BE SUBTLE & CHOSEN TO ADD INTEREST TO THE COLOR SCHEME, NOT COMPETE WITH IT FOR ATTENTION. A DARKER OR LIGHTER VARIANT OF THE BASIC WALL COLOR IS A GOOD CHOICE. ACCENTS SHOULD BE SMALL IN AREA (FOR INSTANCE, THE MOLDING AROUND THE FIREPLACE TILE AS OPPOSED TO AN EN-TIRE DOOR).

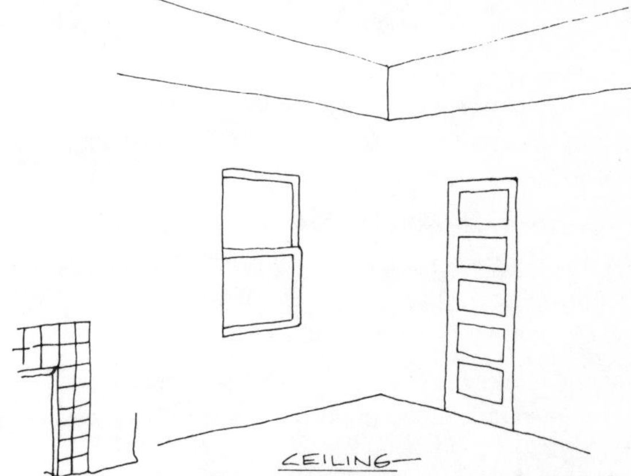

CEILING— PAINT IT WHITE OR A LIGHT EARTH COLOR FOR REFLECTIVITY. DON'T PAINT NATURAL WOOD BEAMS.
DOORS & MOVING WINDOW PARTS SHOULD BE LEFT NATURAL IF THEY AREN'T ALREADY PAINTED. OTHERWISE, PAINT THEM THE SAME COLOR AS THE LOWER WALL AREA OR OFF-WHITE. OFF-WHITE REFLECTS LIGHT INTO A ROOM, & ON DOORS SHOWS OFF THE SCULPTURAL QUALITY OF THE PANELS. DON'T HIGHLIGHT THE DOOR PANELS WITH COLOR; IT LOOKS "SPOTTY." LET SHADOWS DEFINE THE PARTS.
TILE & BRICK SHOULD NOT BE PAINTED. IF IT'S ALREADY PAINTED & BEYOND YOU TO STRIP, PAINT THE BRICK OR TILE AN EARTH TONE THAT COMPLEMENTS THE BASIC WALL COLOR.

water, dries in about 1 hour, and has little or no paint odor. It is available in a flat (matte) finish and a semigloss (satin) finish. An extremely varied selection of colors is another asset.

TYPES

When painting plaster or sheetrock walls and ceilings, use a flat latex. The semigloss latex is best suited for the trim, moldings, and other detailed wood ornaments in areas susceptible to dirt and moisture, such as kitchens and bathrooms. Not only does semigloss provide good texture contrast, but it is easy to maintain because it can be washed. Trim, especially around windows and doors, is the prime fingerpaint and smudge collector of the entire house. Flat latex often comes off, a film at a time, if continually washed or wiped clean; it can be completely wiped off with an ammonia- or bleach-based household cleaner.

Sometimes you want the time- and labor-saving qualities

of latex and the water resistance of the satin finish, but without the gloss. Dutch Boy Paint Company has come up with the solution. They offer a flat-finish latex that has the water-repellent qualities of a semigloss latex enamel. Washable, but not shiny, this paint is ideal for bathroom and kitchen walls and ceiling surfaces.

COLOR SELECTION

Color has an enormous impact upon the way a room looks, feels, and catches and reflects light. Use color to highlight architectural features, not to create a carnival atmosphere. (Refer to the section on color selection in Chapter 9 for an explanation of wise and disastrous color selections and combinations.) Additionally, follow these 13 guidelines regarding interior color design and paint usage:

1. Do *not* paint unpainted wood surfaces, brick or stone, fireplaces, ceramic tile, or floors.

2. Do not use more than two colors in a single room. The safest combination is a light color for the walls and a darker shade of the same color for the wood and trim. Also, a dark, contrasting, earthy color such as a rich brown is dramatically effective when used on the wood detail.

3. Avoid loud or whimsical colors (such as blue, green, red, or even toned-down variations of these colors) for the interior trim.

4. Do *not* use sickly colors like pale olive green, mustard, or chartreuse or garish colors like pink, violet, lavender, aqua, or purple anywhere in a room. Remember, the room is the stage for your furnishings and decorations.

5. Light colors make a room look more spacious. Dark colors make an average-sized room look small, a small room like a cave.

6. Warm colors (yellow, orange, red, brown, warm gray) make rooms with a northern exposure feel more inviting. Both warm and cool colors are appropriate in sunny rooms. However, remember that it is often difficult to successfully combine warm and cool colors.

7. Historically accurate colors are by far the safest selection and can be an excellent conversation topic. This is not to discourage you from a little modern interpretation. Many architectural details and wall surfaces become elegantly "spiced" with flavor when subtly painted to integrate with the style and colors of the furnishings in the room.

8. If at all in doubt, use one color only, preferably white or off-white.

9. When selecting an off-white, pick one tinted with a color compatible with the other colors or woodwork in the room. Hints of blue, green, or brown are really what make the off-white "off."

10. If you are changing colors from room to room, make sure that the rooms blend visually by using colors sympathetic with one another. If in doubt, use the same color scheme throughout.

11. If repainting wainscoting, use earth tones; avoid raw or loud colors and pastels. Browns and beiges present a much more appealing and easy-to-live-with element.

12. Do *not* paint over wallpaper unless it is properly hung and not brittle with age and about to fall off. Do *not* paint over wallpaper with an oil-base paint.

13. Whatever you decide to paint your room, use white on the ceiling because white best reflects light, and a ceiling is one of the surfaces that catches the most light. Thus you will not have to rely on an electric light early in the day. A disadvantage of a white ceiling is that it tends more than any other room feature to show the dirt, especially cigarette smoke and dusty ash from a fireplace. Hot air rises, so dirty films like smoke will stick to the ceiling.

For open areas that present a variety of planes, like a stairway or open-rail balcony, consider the entire picture. Make the elements of the foreplanes integrate with the surfaces behind them. For example, to dramatize a stairway balustrade, use the wall behind it as a foil. If the balustrade is natural wood, paint the wall a light color that matches the rest of the decor. If you are repainting the balustrade, use a color that contrasts yet is compatible with the wall. A dark wall calls for a light balustrade, and vice versa.

Each painting situation varies from house to house and thus has its own unique set of problems. For best instructions regarding required tools other than brushes, scrapers, and rollers, explain your situation to a paint dealer, who usually has encountered one or more obscure circumstances and can correctly suggest specific procedures and equipment or recommend a good consultant.

SURFACE PREPARATION

Remember that, as with exterior painting, wall (and ceiling) surface preparation prior to painting is of the utmost importance. Repair the plasterwork or sheetrock (as discussed in the Plaster section of this chapter), and by all means be sure the wall is clean and free of dust, grease, or particles. Trisodium phosphate (TSP), sold at most hardware and paint stores, is about the best cleanser to use on wall and ceiling surfaces. Follow the label or dealer's instructions for mixing proportions. Do not let the chemical be absorbed into your skin (wear rubber gloves) or get into your eyes. Wash the walls and ceilings with sponge mops, keeping the chemical from contact with your skin. TSP acts like a sponge, sucking up the moisture from your skin and leaving it dry, parched, and painful. The effects to your eyes are even worse.

STRIPPING ENAMEL

If the existing paint is enamel, patching and cleaning the surface is only half the preparation job. You must lightly sand enamel paint to give it some tooth for the new paint to adhere to. Use coarse sandpaper, and sand in a circu-

lar motion. If any fine wood detail that needs repainting has been buried under layers of paint, strip off most of the paint prior to repainting. The stripping job need not be that meticulous because you will be painting it. Just make certain that the definition of detail is visible.

(WALLPAPER)

The first wallpaper color-printing machine was brought to the United States from England in the mid-nineteenth century, and as a result thousands of floral and geometric patterns became commercially available. (The art of wallpaper making peaked in the late nineteenth century.) In most Victorians and in some Colonial Revivals, wallpaper was used to strengthen the composition of the room. The lower walls (below the picture molding and above the wainscoting) usually had a consistent pattern of paper, which was applied in vertical strips, as is the custom today. The frieze area (between the picture molding and the ceiling) usually had a single horizontal band of paper, often in patterns derived from the actual frieze designs in Classical, Renaissance, and Baroque buildings. This treatment gave that portion of the wall above (and including) the picture molding the appearance of a single, huge cornice. Sometimes a small horizontal band of paper was applied directly below the picture molding, giving the molding the appearance of a secondary cornice. Wallpaper was also used on ceilings, in successive strips, as on lower walls, or in borders around the ceiling rim.

Victorian and Colonial Revival patterns varied greatly. Frequently they were used to create lavish effects similar to those created with much more expensive materials like wood, carved plaster, and marble. Interestingly, in contrast to the architectural exuberance typical of the period, many of these wallpaper designs were rather delicate, derived from Art Nouveau or Far Eastern motifs. Authentic Victorian and William Morris wallpaper designs are available at some suppliers.

STENCILING

Wallpaper was often supplemented in the Victorian house by the border effect of stenciling, particularly around the rims of chandelier rosettes, above and below moldings, and within frieze areas. *Stenciling* consists of repeatedly tracing onto a surface a pattern from a template and coloring the pattern with paint. To do this yourself, select a design from a reprinted Victorian pattern book, like the one published by Dover (New York) and make a template from a rigid material like cardboard or plastic. Scale the design correctly, using a gridding system to retain the correct proportion. Transfer the pattern to the cardboard or sheet of plastic and use a matte or fine-blade knife to cut out the pattern. You can repeat the pattern on the wall indefinitely (or until the space is filled up).

SELECTION

In restoration, wallpaper is properly used in consort with the architectural features of a room to add texture and interest. It is often mistakenly used as the focus of a room or as a substitute for valuable woodwork and trim that has been removed. Serious restorationists try to duplicate the precise pattern of the original paper or at least replicate a pattern popular during that era. For some purposes, a conscientiously compatible modern pattern can be quite acceptable as well as economical. Today's "wallpaper" is often made of vinyl, metallic foil, burlap, fabric, cork, and even wood. Using fabric as wall covering, instead of paper, is a good idea if you want a textured, solid color.

Selection of the right wallpaper design starts with color considerations, which are the same as those described for paint. In addition, use the following six guidelines to help select a pattern:

1. The direction of a pattern influences the way a room feels: A strongly vertical pattern makes a low ceiling seem higher, and a dominant horizontal pattern makes a narrow room seem wider. A vertical pattern anchors the wall visually and gives it more solidity than a horizontal or random pattern.

2. The texture of the pattern determines the general impact of any selection. Fine-grained patterns give an impression of overall color, whereas coarse patterns emphasize the body of the individual shapes. On a test swatch of wallpaper the pictograph seems all important, yet on the wall it is the cumulative effect that really matters because the eye tends to blend together the colors and designs. If uncertain, choose a fine-grained pattern with muted colors rather than a large pattern with bold colors.

3. Stylized patterns are more effective than naturalistic patterns because abstracted designs create a textured surface, but realistic representations call attention to themselves, like a painting.

4. For the area above wood wainscoting and for high-ceilinged Victorians, a paper with vertical emphasis complements the architectural features and the interior space. The Victorians had exceptionally busy taste in wallpaper design, but there is no need to abide by their precedent unless it appeals to you.

5. Do *not* use theme papers to try to make a house look "old." Representations of colonial America with horse and buggies do *not* belong in a Colonial Revival House. Likewise, red flocked wallpaper that connotes the Gay Nineties in a pizza parlor is not at all suited to a large Victorian residence.

6. Do *not* use contact paper, which looks shoddy and harms the plaster. There are an ample number of wallpaper products just as convenient: washable, prepasted, scuff-resistant, pretrimmed, and strippable.

Whatever your selection, keep in mind some of the design dos and don'ts of the painting process. The most critical point concerning wallpaper is that of pattern compatibility. If two rooms are visually connected, do not use different wallpaper patterns that are seen from the same point. This looks gaudy and atrociously tasteless. Use the same paper or, better yet, get the most out of the pattern by painting one wall a tint of a compatible color in the wallpaper, to really display the virtue of the design.

(TRIM)

The architectural features are what is left in a room after all the furnishings and portable decorations have been removed. The features stay with the house from one owner to the next; they make a room look complete and contribute to the resale value of the house.

An architectural feature easy to identify is the original wall trim. It is extremely important for design and economic reasons to retain the wall trim. Fortunately, if the trim is missing, it is fairly easy and inexpensive to replace. Trim includes door frames and caps, window frames and caps, cornices, chair rails, baseboards, and assorted molding. Visually, trim breaks up the mass of the wall and adds sculptural interest to an otherwise flat plane. Functionally, it conceals seams and joints.

MOLDING

The words "trim" and "molding" are often used interchangeably because trim is typically made of one or two pieces of molding. In eighteenth- and nineteenth-century American houses, elaborate wood moldings, hand carved with simple tools, constituted the personal signature of the carpenter. By the Victorian period, standardized molding was milled commercially. Eight classic precedents, derived from Greek and Roman designs, constituted the basic shapes for molding and still do today. Victorian builders took advantage of the great variety and convenience of mass-produced molding and combined the patterns into imaginative wall decorations. Although taste became more simplified in subsequent years, molding itself persisted as an essential part of the wall.

Removal. If the molding is still intact, do preserve it. Wall, floor, or ceiling repairs may involve temporary removal of the molding. Remove molding carefully, and only when absolutely necessary. If the molding is painted, sand the surface enough to locate the nailheads. Gently work a stiff, broadblade putty knife behind the molding, lifting it enough to insert a small, flat prybar or the claws of a hammer. Put the end of a prying tool into the gap made by the putty knife, placing it as close as possible to where a nail passes from molding to wall or woodwork. Pad with thick cardboard any spot where the prybar presses on visible woodwork or plaster. Pry the molding away, just far enough to allow the prybar's insertion at the next nail. When the last nail is reached, the whole molding should stand slightly away from the surface of the wall behind it. Now work back from nail to nail, uniformly increasing the gap each time. After two or three passes in this manner, the molding should be loose enough to pull off by hand. Use wood wedges or scraps of shingle to keep the molding from springing back and closing the gap.

Sometimes molding is secured by nails in opposing directions. Check to see if this condition exists before yanking the molding off in one direction, or you will split the molding. If you do split the molding, or if it begins to crack, drive nails with a nailset and hammer all the way

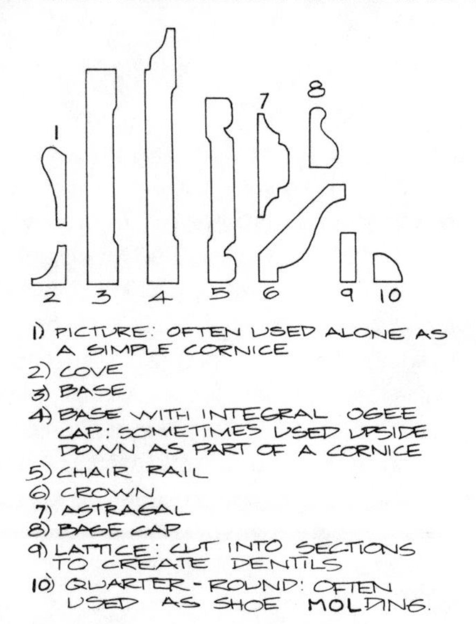

Basic molding shapes

through the molding into the surface behind it. *This procedure works only with finish nails.* You are out of luck if a carpenter used flat-head nails for decorative ornament.

Replacement. Some portion of the molding may be damaged and require replacement, or the molding may be missing as a result of a previous "improvement." There may be a replacement right in the house, for example, if molding was saved when a closet was added or a wall removed. Otherwise, buy some at a lumberyard. You can replace a simple molding easily, but a more complicated pattern requires the combination of two or more basic moldings, just as in Victorian times.

Always buy wood molding, either pine or redwood. Never settle for an extruded metal molding with a wood veneer or a wood molding with a layered vinyl veneer, unless it will be placed so far from view that it will make no difference to the eye. These substitutes are obviously artificial and can ultimately disgrace beautiful furnishings, a restored floor, and a meticulous paint job. If moldings have to be reconstructed, refer to the shapes and styles as illustrated and fabricate them with the same method as for the handrail (see Chapter 8).

If the trim is missing altogether, make a pattern for an exact duplicate of the original molding.

PLAIN, FLAT DOOR & WINDOW TRIM CAN BE ENHANCED BY ADDING CAPS OF THE PROPER PERIOD. THE WINDOW CAP SHOWN WOULD BE SUITED TO A CRAFTSMAN OR COLONIAL REVIVAL WINDOW. BE CONSISTENT: ADD CAPS TO ALL THE DOORS & WINDOWS IN A ROOM AT ONCE.

Adding a window cap

Other Uses. Trim can also be used quite successfully to make a plain house look fancy (as long as it is within the bounds of its architectural character). Supplement plain window or door frames with caps; be certain the caps are consistent with one another. Add a chair rail or a plate rail to the top of the wainscoting, or add cove molding where the wall and ceiling meet. This is not only a visual improvement, but it can really be a great deal of creative fun and an accomplishment. A piece of molding can divide up a space, create lines in which to place bands of color accent, and provide rich value to an otherwise common room or feature. An interesting idea, provided furnishings do not block the greater portion of your work, is to place a piece of molding 4 × 6-inches wide about 3½ inches from the floor on a 9-foot high wall, to create a

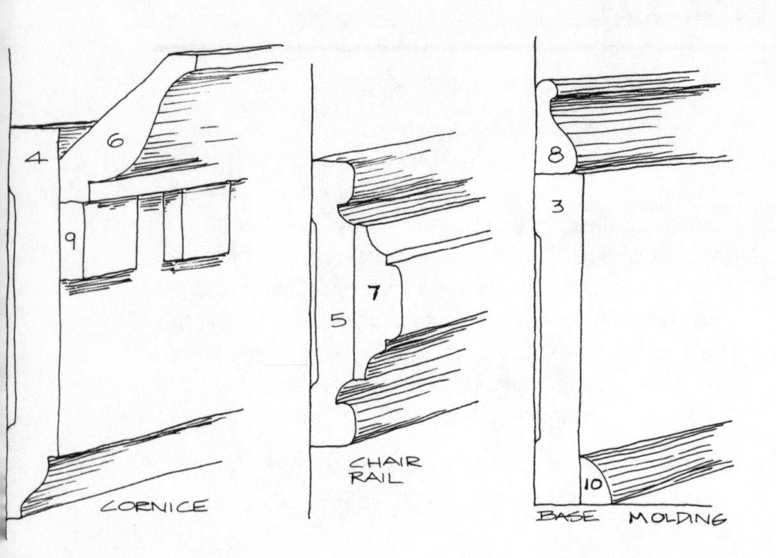

TRIM MAY CONSIST OF A SINGLE MOLDING OR IT MAY BE BUILT UP FROM 2 OR MORE OF THE BASIC SHAPES. THE COMPLEX TRIM PIECES SHOWN ARE NUMBER CODED TO THE PREVIOUS ILLUSTRATION TO REVEAL THEIR INDIVIDUAL COMPONENTS.

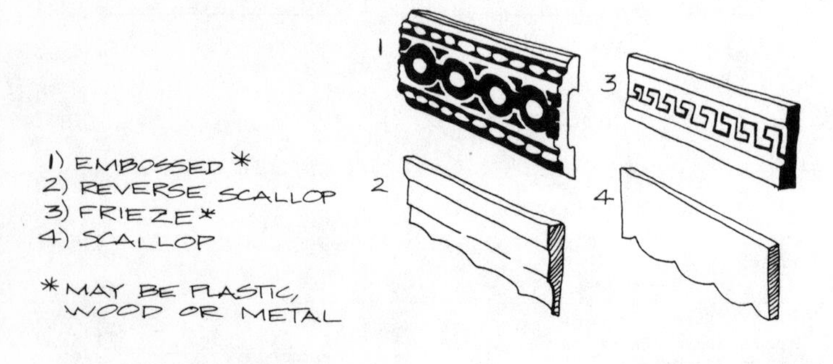

Molding to avoid

dividing line between areas. Paint the lower portion darker than the upper area. If you are stuck with a modern flush door, use molding to imitate the panels of an antique door.

Do not be afraid to be creative. A relatively small investment such as molding strips can often make a visual world of difference in the general appeal of your environment. Just remember to select classic designs that reflect the proportion and character of the house. Do not use scallops, reversed scallops, fake friezes, heavily embossed Spanish styles, or any other decorative pattern that calls attention to itself, unless it is truly compatible with the architecture.

[REMOVING WALLS]

The floor plan of a house can sometimes be improved by removing a wall. For example, a Victorian consisting of a trio of small rooms—one for cooking, one for washing, one for storage—can be converted into a single large room, with a convenient work triangle and a recreation space for the family. In a bungalow, two small bedrooms can be combined into a master suite.

When planning, envision the shape of the space that will result. Keep in mind the proportions of the room, the relationship of height, width, and depth. More floor area is not the only criterion. A room that is made longer must be made wide enough so as not to feel like an alley. Likewise, a room that is made bigger must be high enough so as not to feel like a cellar.

Never remove a wall before its structural role is determined. Load-bearing walls support the ceiling above them. This overhead weight must be transferred to another support before the wall can be removed or the ceiling will collapse. A clue to a wall being load bearing is in the house's foundation. Look at the posts and piers. The intermediate rows they form between the exterior bearing walls are really supports of the weight above them. Any interior wall that rests directly on or along the row of posts below is load bearing. And in many, but not all, cases the exterior walls are also load bearing. Sometimes the foundation is posts and piers, with the main floor girders cantilevered over the post support. The exterior wall and perimeter foundation wall act as a "skin," protecting the basement from the outside and visually anchoring the house to the ground.

Roof lines and ceiling joists can be other clues as to which walls are load bearing. Look at the ceiling joists from inside the attic. Often two joists are crosslapped over a wall; that wall is load bearing. Usually, if the roof has a simple pitch, the ceiling joists run perpendicular to the main ridge beam; the load-bearing walls are those that run in the same direction as the ridge beam.

Definitely consult an architect or engineer, because it is not always completely obvious which walls are load bearing and which ones are not. Do not make any wild guesses yourself. The consequences of removing a load-bearing wall and not compensating for weight distribution can be your house literally caving in. Note too that removal of a wall is not a matter of simple home fix-it work; it requires a building permit. This means that an individual versed in structural statics must redesign the weight-bearing portion so the loads are transferred correctly and construction is safe and secure. A working set of plans must be drawn and approved by the building department. This is valuable insurance that guarantees that the structural stability of your home has been determined safe.

Non-load-bearing walls are really partitions that separate one room from the other. They often look no different than load-bearing walls, except that sometimes they are of skimpier construction. But do not rely on the quality of construction to determine a wall's role—again, get a professional diagnosis.

Before the wall is demolished, salvage any architectural features from it for use in other parts of the house. Molding, wainscoting, doors, door frames, and caps are especially useful for clothing a stripped wall elsewhere.

(CHAPTER 12)

CEILINGS

The height and surface of a ceiling have a powerful influence on the way one feels in a room. For example, a high ceiling suggests formality and provides a setting that seems spacious enough for groups to gather in, even if the dimensions of the floor plan are not that generous. A low ceiling suggests informality and provides a more intimate setting for conversation and privacy. The floor-to-ceiling height should be in proportion to the length and width of the room, to ensure that the interior space will not seem awkwardly squat or elongated. Often the ceiling heights in a house differ from one room to the next, to offer a variety of living environments, but the original ceiling height in each room was intentionally selected to express the function of the room and mood of the architecture. This should be respected in the course of restoration.

Ceilings are prone to cracks, but you should never change the original height to correct this problem. The so-called solution in so many home-repair books to the deterioration in aged plaster ceilings is to install a modern, dropped acoustical-tile surface overhead. This is really no solution at all. It is an unnecessarily costly cosmetic coverup that clashes with the architectural character of the house. Such measures reduce the appeal of the house rather than increase it.

(SURFACE DAMAGE)

Ceiling damage is not impossible to reverse, and the repair process is generally less expensive than the conversion to a modern, inappropriate ceiling style. Plaster ceilings are subject to cracks and holes caused by settlement, water leakage, and improper installation of hooks and fixtures. Always correct the cause of damage, like faulty plumbing, before correcting the symptom. Use the same techniques described in the preceding chapter for the repair of plaster wall, with the additional help of a sturdy ladder or two. Plaster dust from a ceiling can blanket a room, so cover floors and furniture, or remove the furniture entirely.

SHEETROCK

Occasionally plaster ceilings begin to sag because over time the lime in the plaster corrodes the lath nails. The bulging surface can be pressed back into place by drilling a hole through the plaster and installing a ceiling anchor. If the damage is extensive, patch it with sheetrock, in the same manner discussed in surface wall repair in the last chapter. Always attach sheetrock with special sheetrock nails, and anchor the piece to the ceiling joist, not to lath that has been left on. As with walls, when applying sheetrock, make a slight dent in the surface when nailing. Use a waffle hammer to impress some tooth into the soft sheetrock, and spackle over the little craters when finished, just as you do when you tape seams. This prevents the future appearance of protruding nails caused by settlement and just plain aging.

MODERN TILE

Compare the cost of sheetrocking the ceiling of a small (say, 8 x 12 foot) room with the cost of resurfacing the ceiling with acoustical tile. The sheetrock method costs about $30, complete (three sheetrock panels, 3 x 8 foot x ½ inch at $4 a sheet; sheetrock tape and joint compounds at $10; one hi-pile paint roller with a 9-inch frame at $5; one or two paint brushes for $3). The acoustic tile method costs $77, more than twice as much (96 square feet of acoustic tile at 55¢ a square foot costs $52.80; 96 linear feet of 1 × 3-inch furring strips at 20¢ a linear foot costs $19.20; staple gun and staples for $5). The larger the ceiling, the greater the price differential.

If you *must* use a modern ceiling surface, follow these suggestions:

1. Select a product that is plain white, without patterns, glitter, or volcano eruptions. Respect the fact that the original plaster was flat and plain.

Ceiling: do's and dont's

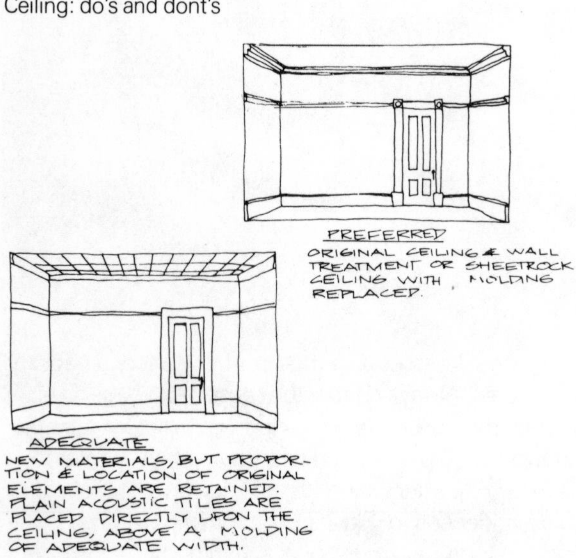

PREFERRED
ORIGINAL CEILING & WALL TREATMENT OR SHEETROCK CEILING WITH MOLDING REPLACED.
ADEQUATE
NEW MATERIALS, BUT PROPORTION & LOCATION OF ORIGINAL ELEMENTS ARE RETAINED. PLAIN ACOUSTIC TILES ARE PLACED DIRECTLY UPON THE CEILING, ABOVE A MOLDING OF ADEQUATE WIDTH.

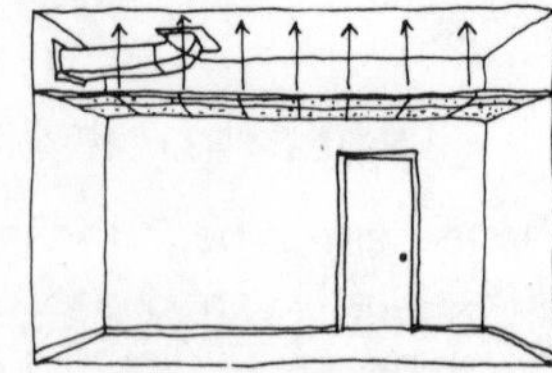

OBJECTIONABLE
LOWERED CEILING WITH HOLES OR HEAVY TEXTURE ON THE SURFACE OF THE PANELS. DECORATIVE ELEMENTS AND PROPORTIONS OF THE ROOM ARE DISREGARDED.

2. Select as neutral a surface as possible. The ceiling is not a planetarium.

3. Install the material at the same level as the original ceiling. If the plaster is in place, smooth it out by the same techniques described for plaster repairs, and glue the tiles directly to the plaster with a special contact adhesive manufactured for that purpose. If the plaster and lath have been removed, nail furring strips no more than 1 inch thick direcly to the joists of rafters and staple the tiles to them.

(HEIGHT)

Never suspend the ceiling. Dropped ceilings are often used for questionable cosmetic concealment; as modernization; to hide new wiring or plumbing; as acoustic mufflers; and for heat conservation. However, there are alternative approaches to each situation that are much more sensitive to the architectural style.

To minimize heat transfer and thus reduce fuel costs, install insulation between the interior wall surface and the exterior wall. If the wall is to be rebuilt, blanket insulation is the typical selection. If the walls are in place, an installation specialist can blow loose fill insulation into the space between the walls. To muffle sounds, use rugs, drapery, upholstery, or fabric wallpaper to absorb the reverberation. To isolate sounds from other rooms, use two layers of wallboard instead of just one, if the base molding and other trim will accommodate this. To conceal new plumbing and wiring, install modernized utility systems so that closets, shafts, and floor space can be used to their best advantage. (Refer to Part 4, Utility Systems.)

(SPECIAL FEATURES)

Always retain special architectural features that distinguish a ceiling. If the features are missing, but belong to the style, it is to the building's advantage to replace them.

ROSETTE

Victorian and Colonial Revival ceilings had a highly decorative rosette in the middle, as a support for the chandelier. A *rosette* is a circular plaster sculpture shaped like a flower. Small rosettes were moleded into a single piece as small as 6 inches in diameter; others were

Rosette cross section

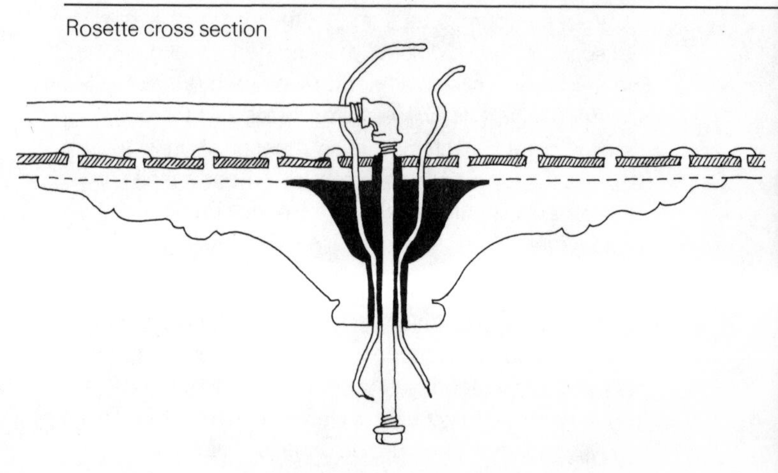

THIS IS A CROSS SECTION OF A PLASTER ROSETTE. APPLIED TO THE CEILING WITH PLASTER OF PARIS (AND WIRE TIES FOR THE LARGE SEGMENTED ROSETTES). A HOLE WAS PROVIDED FOR A GAS PIPE & THE CAPPED PIPE OFTEN REMAINS DESPITE THE REPLACEMENT OF GAS FIXTURES WITH ELECTRIC ONES. ELECTRIC WIRES WERE RUN THROUGH THE PIPE HOLE.

assembled in several sections and were as wide as 4 feet in diameter. The rosette design and size were in proportion to the size of the ceiling, the volume of the room, and the shape of the original light fixture.

In Victorian times, the rosette was integrated into the ceiling plaster. Segmented rosettes were additionally supported by tie wires, threaded through the lath to the structural member above. Thus major repairs to the rosette require first cutting it out of the ceiling. Saw through the lath around the perimeter of the rosette, supporting it as you do so. Use a long chisel to loosen the lath from behind and a hacksaw to sever the tie wires. This may be easier to do from the floor or attic above the ceiling.

Once the rosette is off, strip accumulated paint by dipping the rosette in a pan of very hot water on the stove and scrubbing it with a toothbrush. The difficult layer is the calcimine, the layer closest to the plaster. Fill in small cracks and holes with plaster. If necessary, rebuild missing sections with a self-mold and casting plaster. When sanding, be careful not to smooth sharp edges or you will destroy the detail, instead of resurrecting it. Prime and paint, as you would a new rosette.

Reinstall the rosette in the original way by replacing it in the hole in the ceiling and securing it with tie wires and a fresh coat of plaster; or secure the rosette with bolts and screws to the joists.

If the rosette is missing, broken, or obscured by countless layers of paint, you can buy a new one. Certain distributors carry rosettes in a variety of styles, with detailed installation instructions.

Even if the original light fixture is missing, you can still take advantage of the rosette. Install a pipe-type chandelier that looks appropriate, even if it is not an accurate reproduction. (Refer to Chapter 14, Electrical System, for installation guidelines.) If the gas line has not yet been sealed, have a qualified plumber cap off the line. Amateurs should *never* handle gas lines that may still be active.

TRIM

Many older homes have wood ceiling moldings, and a common Victorian molding was made of plaster. Retain or replace ceiling molding and cornices. Half a century or more of layers of paint can result in a nasty accumulation that blurs the crisp juncture of the ceiling and wall. With a manual paint scraper, score and remove the buildup. Sand the edge of the newly exposed layers of paint so that the transition from thick paint to scraped corner is as smooth as possible. To conceal rather than correct the buildup, paint the corner as is, then add a stip of suitable cover molding around the perimeter and cover it up. This useful solution saves time, and for a relatively minimal cost it adds a richly finished appearance to the room.

BEAMS

Some older homes boasted exposed wood beams in the living room and dining room ceilings. Other homes had beams that imitated *coffered ceilings* (square pockets gridded by the beams). If wood ceiling ornamentation has been painted so many times that the angular corners have become rounded, or if you want to reveal the natural wood (as so well suited to the Classic styles), strip the paint by one of the stripping techniques described in Chapter 11. Natural wood beams are dramatic and a real architectural asset because they are rarely used in modern residential construction.

COVED CORNERS

Coved ceilings are found in selected Victorian houses and some examples of the Colonial Revival and Period Revival styles. Coved ceilings were created by attaching lath strips above the cornice or picture molding to a curved framework instead of to the studs. This framework usually describes a quarter circle between the wall

Coved ceiling detail

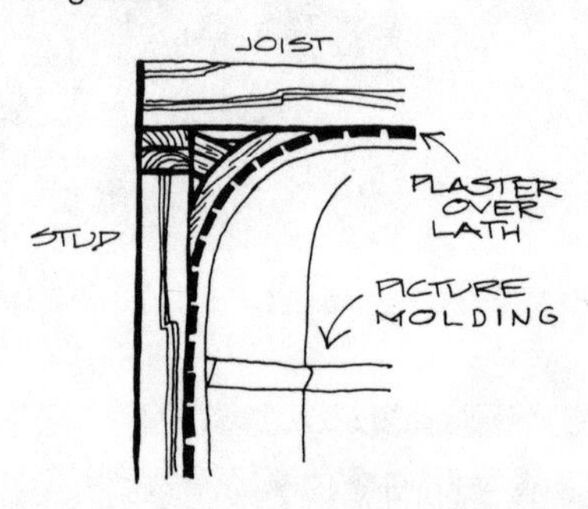

and the ceiling surfaces. Repair techniques for a coved ceiling are identical with those for conventional plaster walls. However, if damage is severe enough to warrant replacement, you must replaster the surface. Sheetrock cannot be used because it is rigid and cannot be nailed to a curved surface. If coved corners are intact, keep them in place. If they are damaged, repair them, because coved corners give a room a special sense of enclosure that makes the space particularly nice to be in.

(CHAPTER 13)

STAIRCASES AND DOORS

A staircase consumes much space inside the house and can be seen from many angles. Its structure and ornamentation are unique within the house. When improvements are made to it, you should respect and retain the original combination of function and decoration.

(STYLE)

In two-story Italianate and San Francisco Stick houses, the staircase is straight and narrow, running the entire length of the entry hall. To maintain the tall, thin proportions and to traverse the considerable distance from first to second floor may seem awesome because the treads are rather low and the risers quite high. In the large Queen Anne house, the entry hall merged with the living area, so the stairs became a focal point. The hall was square, and the direction of the steps doubled back at a landing midway up the flight. Victorian stairways were distinguished by turned balusters, one or two per tread; carved spandrels; and substantial newel posts. Often the partition wall concealing the *stringers* (horizontal timbers that connect uprights) was surfaced with Lincrusta-Walton.

During the Colonial Revival era, the landing at the kink of the staircase was elaborated upon, with built-in benches, hidden storage, and stained glass windows. The balustrade, simpler than that in the Victorian and more classical in detail, was often, but not exclusively, painted. The staircase was faced with wainscoting, like the other walls within view.

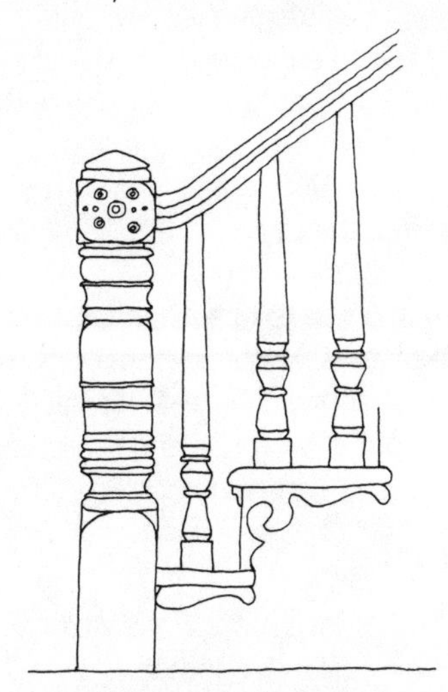
NEWELL POSTS ARE THE OFTEN HEAVILY ORNAMENTAL POSTS THAT ANCHOR THE ENDS OF THE HANDRAIL. SPANDREL ORNAMENTS ADORN THE AREA UNDER THE TREAD END TRIM.

Newell post

(REPAIR)

When restoring the staircase, it is important to differentiate between repairing and rebuilding. *Repair* refers to general maintenance, like firming up loose parts; replacing treads and risers here and there; readjusting the rail and newel posts or replacing them in kind; and patching, puttying, and other fix-it jobs that have no real bearing on structural integrity. *Rebuilding* constitutes an actual reconstruction of the support system and other essential elements.

Normally, if reconstruction is required, the entire staircase

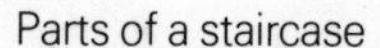
Parts of a staircase

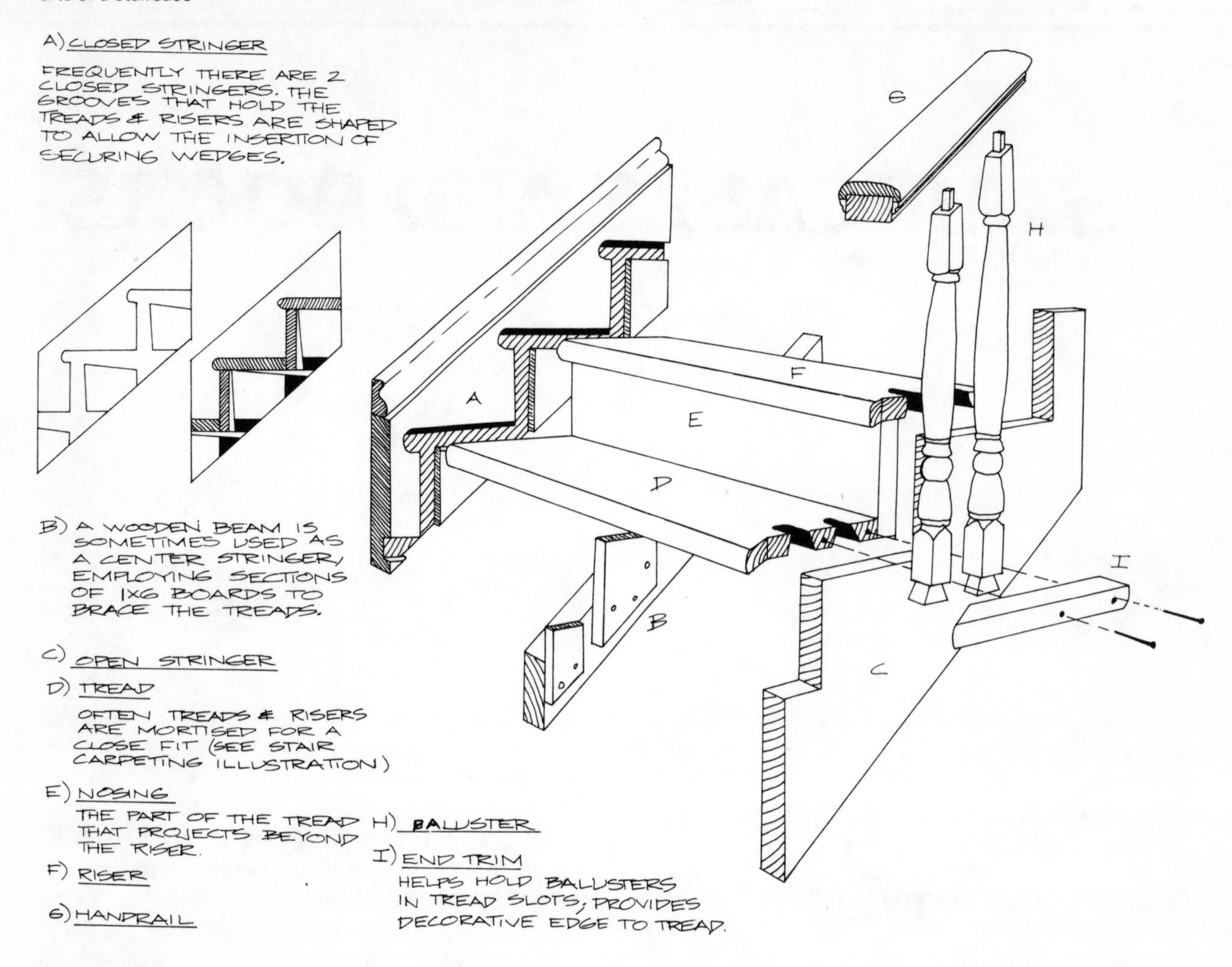

must be rebuilt in accordance with modern codes, which establish stair standards quite different from those of the historical styles. However, in the state of California, the building code has been amended to permit owners of single-family and duplex dwellings to rebuild an historical staircase according to its original design. But the law does *not* permit any variance from the original design and its dimensions. If the staircase is to duplicate its original look, it must do so faithfully, or it has to conform to present standards. For example, a staircase that originally had 9-inch risers could be reconstructed with 9-inch risers, even though today's code requires a maximum of 8-inch risers for new construction. But you would not be permitted to rebuild the same staircase with 8½-inch risers, which is neither the original size nor in conformance with code.

A) CLOSED STRINGER
FREQUENTLY THERE ARE 2 CLOSED STRINGERS. THE GROOVES THAT HOLD THE TREADS & RISERS ARE SHAPED TO ALLOW THE INSERTION OF SECURING WEDGES.
G
H
A
F
E
D
I
B
C
B) A WOODEN BEAM IS SOMETIMES USED AS A CENTER STRINGER, EMPLOYING SECTIONS OF 1X6 BOARDS TO BRACE THE TREADS.
C) OPEN STRINGER
D) TREAD
OFTEN TREADS & RISERS ARE MORTISED FOR A CLOSE FIT (SEE STAIR CARPETING ILLUSTRATION)
E) NOSING
THE PART OF THE TREAD THAT PROJECTS BEYOND THE RISER.
F) RISER
G) HANDRAIL
H) BALUSTER
I) END TRIM
HELPS HOLD BALUSTERS IN TREAD SLOTS; PROVIDES DECORATIVE EDGE TO TREAD.

Rebuilding a staircase is a sophisticated job for an amateur, and (as in the removal or reconstruction of walls) it requires working plans that must be submitted to the local building inspection department for approval prior to any construction. Before taking down your staircase, make sure you have these plans already drawn, submitted, and approved.

If the stair stringers need replacing, chances are your entire stairway must be dismantled. When doing so, be especially careful to preserve the valuable components, such as the balustrade, handrail, wainscoting, and spandrel. Reuse them; it will save both money and the appeal of the stairway. Because you will be completely dismantling the structure, you are legally classified as rebuilding it, so either replace the stringer in kind (usually a redwood 2 × 4 or 3 × 4) or hire a contractor or architect to draw the plans. Be sure the treads, risers, and railing height also comply to code. Fortunately, the interior staircases of the Victorian and similar eras had rails of heights that either meet or can easily be altered to meet today's requirements. (That is why saving the balustrade is a good idea.)

If only some of the surface parts of the stairs need repair, the whole structure does not need rebuilding. The balustrade is a likely spot for problems; it may be wobbly, damaged, or missing altogether. The newel post is the anchor of the balustrade at each story and the landing, so it must be securely attached for the balustrade to be sturdy.

To tighten a loose baluster, drive wedges into the gap

STAIRCASE AND DOORS

between it and the handrail. Begin with a strip of hardwood slightly larger than the gap. Sand or plane it into a wedge, with the grain running the length of the wedge. When the size is right for a snug fit, coat the strip with glue and tap it into position. Another technique is to drill a screw hole through the post at an upward angle toward the rail. Countersink the screw.

A baluster can be removed to strip the paint or for use as a replacement pattern. Typically, the bottom of the baluster fits into a slot in the tread; the top fits into a hole on the underside of the railing. Pry off the end trim on the tread to slide the baluster out of its slot. Then pull down to free the top from the handrail.

To reproduce missing turned balusters, you need a lathe; or you can buy a reproduction of a baluster. One or two pieces are not very expensive, but if most of the balusters are missing, replacing them can cost a lot. You can also duplicate the silhouette shape or even just the proportion of the balusters (refer to Chapter 8, the section on wood porches). *Never* replace an original wood balustrade with one made of prefabricated wrought iron or another material totally out of character with the design.

If the balustrade is natural wood, leave the grain exposed. If it is already painted and needs a new coat, consult the color selection section in Chapter 11. Before repainting, you may have to scrape or strip at least the surface layers of accumulated paint to uncover the real form of the wood elements.

When surfacing the staircase, the intriguing sequence of right angles that epitomizes the structure should be shown off. The obvious way is to leave the wood exposed, but even if the stairs are to be carpeted, the configuration of stair and nosing need not be obliterated. Select a carpeting with a short, dense pile and a minimum of padding, and contour the carpeting to the shape and angles of the step.

Staircase carpeting: do's and dont's

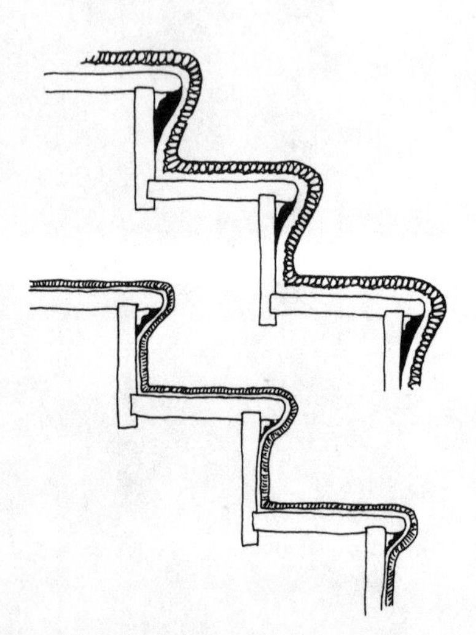

SHAG OR DEEP-PILE CARPETING COUPLED WITH A THICK PAD (TOP SKETCH) CAUSES A LUMPY, OVERSTUFFED LOOK OUT OF CHARACTER WITH EARLY STAIRCASES DESIGNED FOR RUNNERS. THINNER BUT DENSER CARPETING WILL STILL WEAR WELL (LOWER SKETCH) WHILE RESTORING VISUAL DEFINITION TO THE NOSING.

(DOORS)

STYLE

There are three types of interior doors, as classified by construction method: panel, hollow core, and solid core. The three-dimensional panel door (see illustration) is original to the various historical styles discussed in this book. It is made of a wood outer frame that is inset with recessed wood panels or clear, frosted, stained, or leaded glass.

In sorry contrast, the hollow-core and solid-core doors are plain, flat, and mass produced. The hollow-core door is

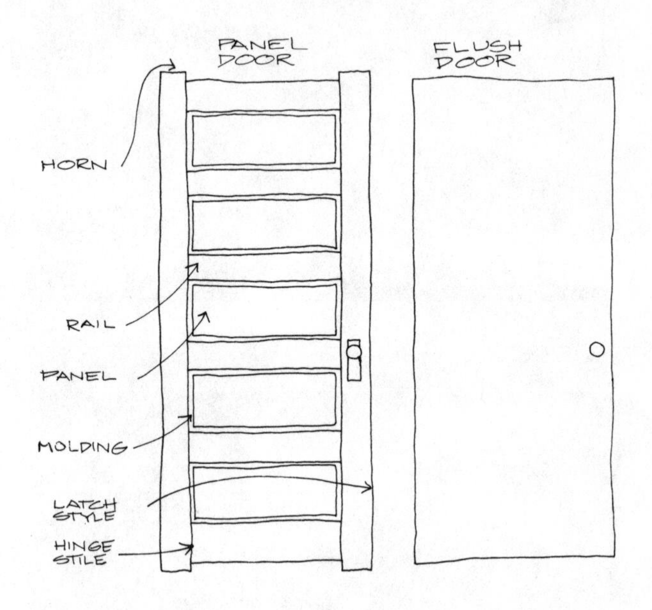

Types of doors

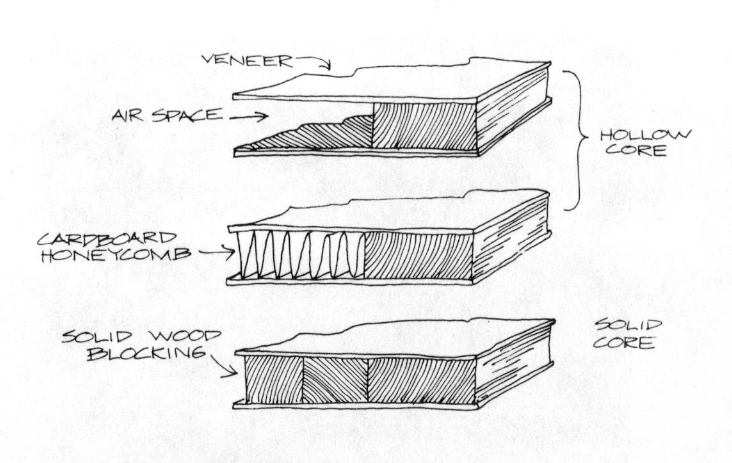

Flush door construction

made of a wood frame filled with honeycomb cardboard and covered by sheets of lightweight veneer. The solid-core door is made of layers of wood laminated with glue.

It is a grievous error to replace panel doors at the entry to a room or a closet with solid- or hollow-core doors or to camouflage a panel door under a sheet of plywood. Here is why:

1. Panel doors match the older architectural features in rooms; core doors do not. The molding that outlines the recesses, the proportion of the panels, the sturdiness of

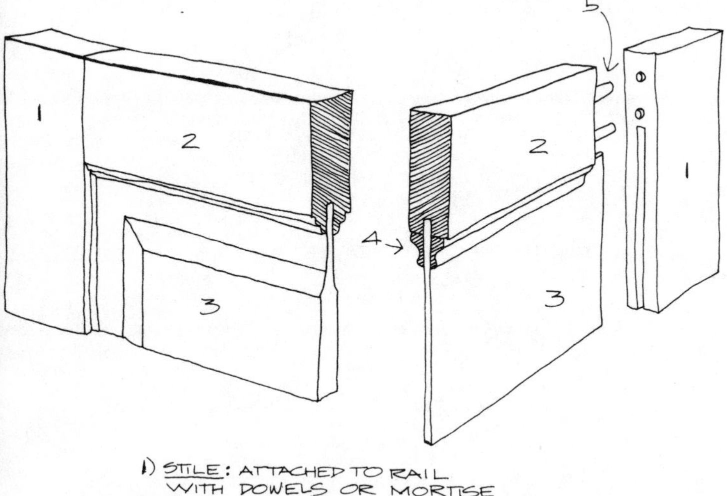

Panel door construction

construction, and the solidity of appearance all look right with the other architectural features in the room. A core door does not pick up on any of the detail and is therefore woefully out of place.

2. Panel doors match the door frame, which is a blend of molding, trim, cap, and base. In this context a core door looks like the piece of cardboard in a fancy picture frame before the work of art has been mounted.

3. Replacing a worn panel door with a modern door is really wasteful. Most problems with panel doors can be repaired at practically no expense, resulting in a door that is far more durable than a core door—particularly a flimsy hollow-core door—will ever be.

4. Because of their solidity, panel doors offer more sound insulation than hollow-core doors.

5. Panel doors are usually outfitted with more attractive hardware; the knobs and hinges for core doors cannot compare in detail.

6. Panel doors are architectural features that stay with the house and add to its resale value, contrary to what a core door salesperson may tell you.

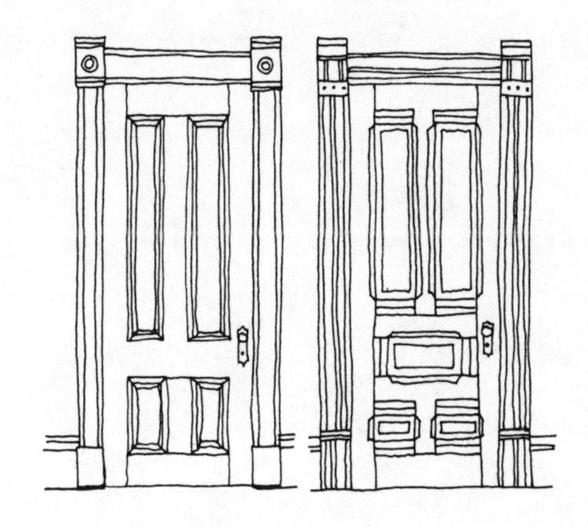

VICTORIAN DOORS ARE PANEL DOORS WITH A VERTICAL EMPHASIS. THEY DISPLAY SUCH A VARIETY OF MOLDINGS & EMBELLISHMENTS THAT IT IS HARD TO ASSIGN SPECIFIC DOORS & DOOR FRAMES TO EACH RECOGNIZED VICTORIAN ARCHITECTURAL STYLE.

DOORS WITH VERTICAL PANELS WANED IN POPULARITY AFTER THE VICTORIAN ERA, & THE HORIZONTAL PANEL DOOR, COMMONLY WITH FIVE PANELS, WAS THE DOMINANT FORMAT DURING THE CRAFTSMAN PERIOD. THE FIVE-PANEL DOOR YIELDED TO THE SINGLE-PANEL DOOR ABOUT 1915. RUSTIC DOORS OF VERTICAL BOARDS, FOUND IN PERIOD REVIVAL HOUSES, WERE AN EXCEPTION TO THE PANEL DOOR TREND, WHICH ENDED WITH THE PROLIFERATION OF FLUSH DOORS IN THE 1940S & 1950S.

Victorian doors

FIT

All doors are subject to binding or looseness, but panel doors more so, simply because of their years in service. Usually this is not a fault of the door itself but poor fit at the hinges and lockset or excessive layers of paint. Correcting a balky panel door is so inexpensive that there is no justification for switching to a modern door style.

The three common causes of balky doors are: (1) a set of improperly adjusted hinges, (2) a distorted door that no longer fits the frame, and (3) a distorted frame that no longer fits the door. To determine the cause, stand so the door opens toward you, but keep the door closed. Take a coin or a piece of wood $\frac{1}{16}$ inch thick and slide it along the space between the door and the jamb. Pinpint the sticky spots where the coin will not slide. If there is a large gap at the opposite side of the door from the stubborn spot, the problem is a loose hinge. If the door fits too tightly all the way around, and you cannot fit the coin on the top and bottom, the door has probably swelled from dampness or paint; the troublesome edges require planing. If the coin hits several sticky spots at random locations, and if there are cracks in the plaster around the door frame, the house has probably settled, causing the door frame to shift position and press on the door at

Typical problems on interior doors

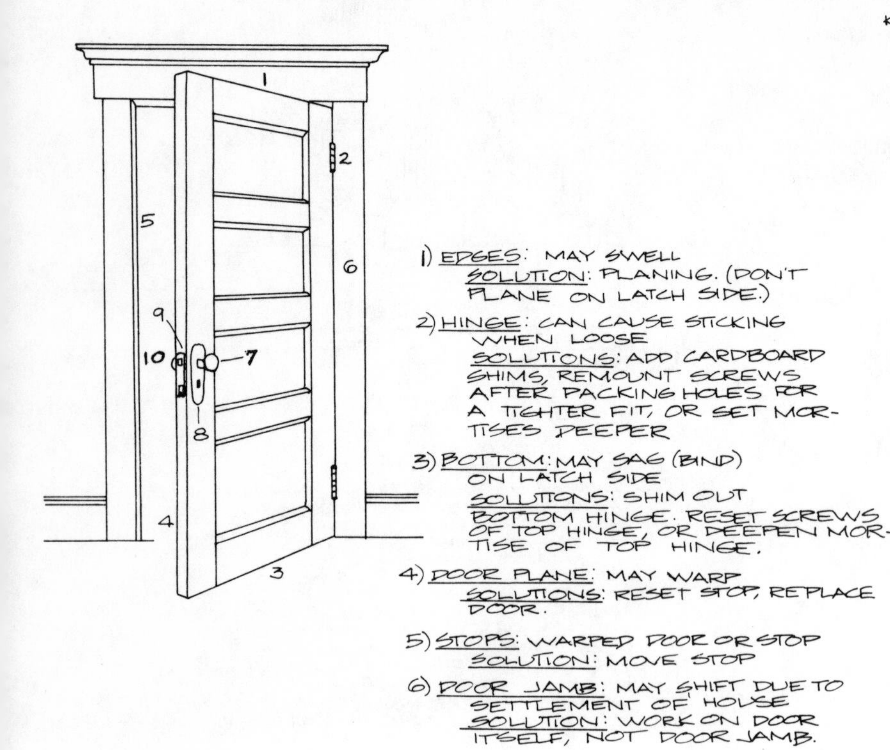

1
2
3
4
5
6
7
8
9
10
1) EDGES: MAY SWELL
SOLUTION: PLANING. (DON'T PLANE ON LATCH SIDE.)
2) HINGE: CAN CAUSE STICKING WHEN LOOSE
SOLUTIONS: ADD CARDBOARD SHIMS, REMOUNT SCREWS AFTER PACKING HOLES FOR A TIGHTER FIT, OR SET MORTISES DEEPER
3) BOTTOM: MAY SAG (BIND) ON LATCH SIDE
SOLUTIONS: SHIM OUT BOTTOM HINGE. RESET SCREWS OF TOP HINGE, OR DEEPEN MORTISE OF TOP HINGE.
4) DOOR PLANE: MAY WARP
SOLUTIONS: RESET STOP, REPLACE DOOR.
5) STOPS: WARPED DOOR OR STOP
SOLUTION: MOVE STOP
6) DOOR JAMB: MAY SHIFT DUE TO SETTLEMENT OF HOUSE
SOLUTION: WORK ON DOOR ITSELF, NOT DOOR JAMB.

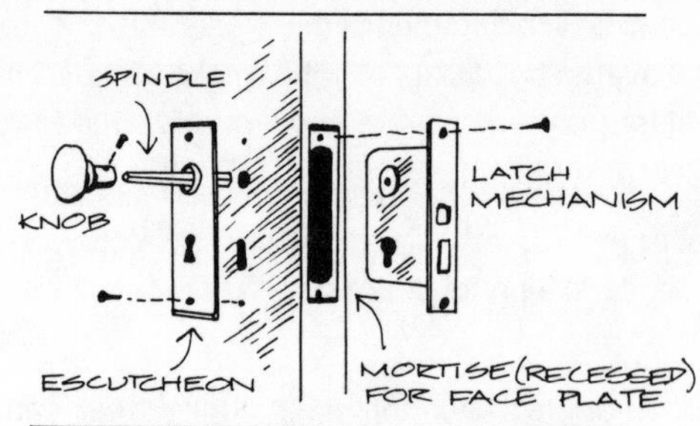

7) KNOBS: KNOB MISSING OR REPLACED WITH ONE INAPPROPRIATE TO DOOR & HOUSE
SOLUTION: FIND KNOBS MATCHING OR IN CHARACTER WITH ORIGINALS; OR USE KNOBS FROM LESS CONSPICUOUS LOCATIONS SUCH AS CLOSETS.
IF KNOB IS JUST LOOSE: TIGHTEN MACHINE SCREW THAT HOLDS KNOBS TO SPINDLE. (OBTAIN REPLACEMENT IF SCREW IS STRIPPED OR MISSING.)

8) ESCUTCHEON: MAY BE TARNISHED OR PAINTED. (ALSO, HINGES, KNOBS, FACE PLATES, & STRIKE PLATES)
SOLUTION: USE PAINT REMOVER. (TAKE OFF DOOR FIRST) & BRASS POLISH. REPLATE IF NEEDED.

9) LATCH MECHANISM: CAN HAVE MECHANICAL MALFUNCTIONS
SOLUTIONS: CLEAN & OIL. IF THIS FAILS, CONSULT A LOCKSMITH OR FIND AN OLD LATCH THE SAME SIZE THAT WORKS.

10) STRIKE PLATE: MAY BE OUT OF ALIGNMENT WITH LATCH, PREVENTING PROPER CLOSURE
SOLUTION: ENLARGE MORTISE ENOUGH TO MOVE STRIKE PLATE UP OR DOWN AS NEEDED.

several points. This too requires planing. It is far simpler to fix the door than the door frame. Planing can be done manually and involves no more skill than a "feel" as to where the sticking is occurring — and this should be obvious. After planing away the trouble spot, sand down the edge to a nice smooth surface before finishing it.

DAMAGE

Worn panel doors may have a kicked-in panel or a loose rail. Thanks to its construction, the panel door can be dismantled and the damaged piece repaired or replaced, as illustrated. Hollow- or solid-core doors, once damaged, cannot be repaired; they have to be replaced.

Often, after a succession of occupants over the course of half a century, paneled doors become swaddled in layers of paint. The crisp edges and recesses lose their definition, and the angular corners become rounded. To resurrect the geometric form of the door, strip the paint, even if you plan to paint it again. (Refer to the instructions for stripping woodwork in Chapter 11.)

Use a semigloss latex enamel to repaint a painted door. (Follow the color-selection guidelines in Chapter 11.) Remember, for enamel-type finishes to adhere properly, especially to undercoats of enamel, sand the surface a bit to provide some tooth for the new paint to stick to and to spread over evenly. Fill nicks and gouges with wood putty.

If the panel door is missing or damaged beyond repair, replace it with a door that suits the style of the house, as indicated by panel doors or trim designs still in place. You can find such doors at specific suppliers or even at salvage yards if you have the patience. If you cannot afford a new door, switch one of your existing doors. Sometimes you never close a certain door, so it is really unnecessary where it is. If this "superfluous" door fits the frame of the missing door, you are in luck. Even if it misfits a

Panel door repair

- REMOVE DOOR
- CAREFULLY, WORK A PRYBAR BETWEEN TOP RAIL & STILE. PRY OFF STILE GRADUALLY, WORKING DOWN ALONG THE STILE AT THE POINTS WHERE IT MEETS THE RAILS.
- PRY OFF TOP OR BOTTOM RAIL IF TOP OR BOTTOM PANELS ARE DAMAGED. OTHERWISE REMOVE SECOND STILE ENTIRELY.

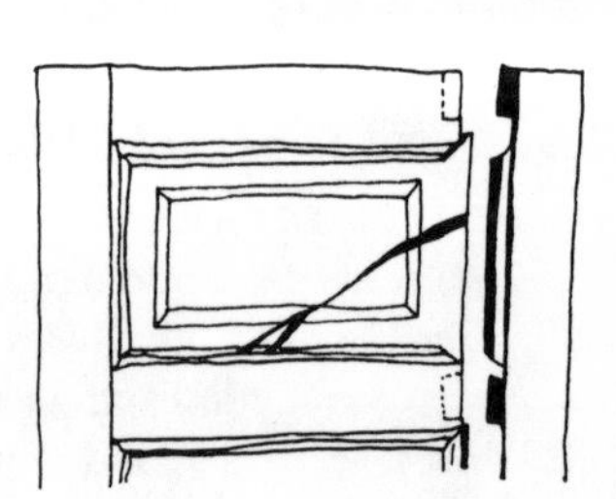

- REMOVE DAMAGED PANEL. REPLACE WITH MATCHING PANEL FROM SALVAGE YARD OR LESS PROMINENT DOOR. IF THE PANEL IS FLAT, 1/4" PLYWOOD MAY DO AS A REPLACEMENT.
- CLEAN DOWELS OR TENONS & APPLY A THIN COATING OF WOODGLUE. REASSEMBLE DOOR, ALLOWING GLUE TO DRY BEFORE HANGING. FASTEN ANY LOOSE MOLDINGS.

little, plane it to fit correctly. If it grossly misfits, see if you can play musical doors around the house — find the closest fit for every door. Measure the doors and frames first to save yourself the work of much unhinging and rehinging.

DOOR FRAME

The door frame is as much a part of the doorway as the door itself. The style of the trim goes with the design of the door, so it should always be left in its original form and repaired or replaced as necessary. Door frames, like other types of architectural trim, are constructed of components. The Victorian examples are often quite elaborate, the Colonial Revival style refined, and the Classic straightforward. Often the design of the door and the frame has been held over from the preceding architectural styles, so frequently you find Victorian doors and frames on Colonial Revival houses and Colonial Revival doors and frames in Classic houses.

The door frame is a good place for embellishment. You can make a simple doorway much more interesting by adding a door cap on the head trim or an extra strip of molding on the jamb. Make sure the embellishments are in keeping with the style of the room's trim and the general character of the house. For example, on a Victorian house use the exterior window ornament as a guideline. (Refer to Chapter 11 for more information about sources of molding and creative ways to combine standard parts.)

(PART FOUR)

UTILITY SYSTEMS

Electricity, heating, and plumbing compose the central nervous system of the house. Because they are subject to continual use, older systems are susceptible to breakdowns, and they become outmoded by modern conveniences. This section is not a complete instruction guide to the repair and re-design of the house's infrastructure. You should consult licensed engineers, technicians, and contractors regarding the specific requirements of the National Electric Code, Uniform Plumbing Code, local ordinances, and so on. However, this section does provide basic information to help you evaluate the present systems and improve them without disastrously affecting the architecture.

(CHAPTER 14)

ELECTRICAL SYSTEMS

Identifying your electrical service

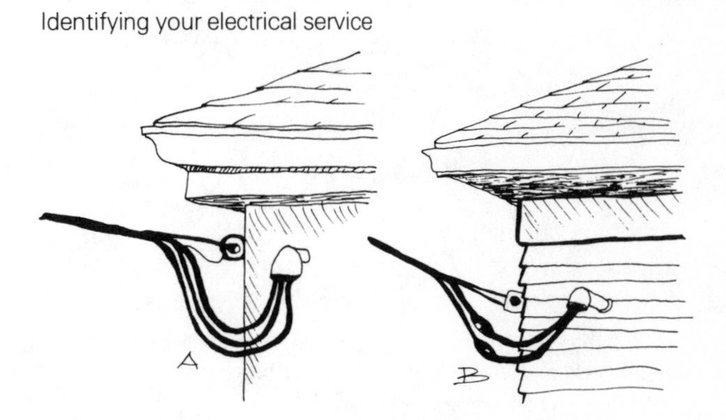

A BRIEF LOOK AT THE ELECTRIC POWER HOOKUP WHERE IT ENTERS THE HOUSE CAN TELL YOU THE VOLTAGE OF YOUR SERVICE, BUT NOT THE AMPERAGE (AMOUNT OF CURRENT) AVAILABLE. A 3-WIRE HOOKUP (SKETCH A) INDICATES 110-220 VOLT SERVICE. THE AMPERAGE MAY VARY FROM 30-100 AMPS. 2 WIRES (SKETCH B) MEANS 110 VOLT SERVICE THAT MAY BE 20 OR 30 AMPS. WIRE SIZE IS THE KEY TO AMPERAGE. FUSE SIZE & FUSE PANEL LABELS CAN BE MISLEADING. ASK AN INSPECTOR OR QUALIFIED ELECTRICIAN.

To begin to understand the way the electrical system in your house is put together and to recognize its problems, you need the following basic vocabulary.

(TERMINOLOGY)

Electricity is moved through wires by pressure, much the way water is moved through pipes. The amount of pressure is called *voltage* and is measured in units called *volts*. *Current* is the rate at which electricity is delivered—to the house from the power pole and to an individual appliance from the wall socket—and is measured in amperes, commonly called *amps*. The voltage forces the current along the electrical wires. The amount of current that can be transported is determined by the diameter of the wire.

The route of the wiring is called the *circuit*. Typically, several fixtures and sockets share a circuit, whereas heavy-duty appliances like refrigerators or washing machines should each have their own circuit. Each circuit is protected from overload by its own *fuse* or *circuit breaker*. There are several fuses, as many as needed for the circuits. These are usually located in a place inside the house (sometimes a closet or enclosed porch) and set into the wall in a *fuse box*. Fuse boxes are typical of older homes. Circuit breakers, the modern replacements for fuses, are commonly found in newer homes, often attached to the exterior of the house. Do not be shocked if you find a *breaker box* in your older home; be pleased, because it indicates that the house has been rewired, and thus many problems from aging wiring should not occur.

It is easy to distinguish fuses from circuit breakers. The fuse contains a strip of soft metal that melts from excessive current. The circuit breaker has a spring that is activated when excessive current passes through it; the circuit is broken as if a switch has been turned off.

(PROBLEMS)

A common problem, especially when many appliances are being used (such as during a busy day of home restoration), is for the power to shut off without notice. This is a working example of a *circuit overload.* Too much current is flowing through the circuit, which causes your fuse to "blow." You must install a new fuse. Just go to your fuse box, find the dead fuse (denoted by its being blackened or the metal melted and by its warmth), and turn the switches to off or you could get fried. Remove the old fuse by turning it in a counterclockwise direction. Replace with a new fuse of the same amperage by screwing it in a clockwise direction.

To prevent fuses from blowing, as they inevitably will if you continue to overload the circuits, run heavy-duty appliances and tools on separate circuits, so that each gets a fair share of current. A good idea for present and future use is to map out the circuits of your house. Draw a simple floor plan, locating on it each outlet and switch. Then pull out all your fuses, except one, and activate all the outlets and switches in the house. Only those controlled by the fuse in the box will be in operation, so you now know which outlets and switches are controlled by this fuse. Mark this down on your floor plan with either a number or a color. Label the spot next to the fuse in the fuse box with the same identification. Follow this procedure for all the circuits and fuses, and keep the floor plan near the fuse box or somewhere handy. When power shuts off for no apparent reason, or when electrical repair work is to be done, then you can readily correct problems.

[SERVICE]

Service refers to the total amount of voltage and amperage available to a house and the number of wires that carry the electrical power from the transmission pole to the meter box on the building. Most older houses (those built before World War II) originally had two-wire, 30-amp, 12-volt service. This was perfectly adequate for electrical needs then, and may still be for those of you with Spartan energy needs; but two-wire, 30-amp service is very limited for today's appliance-reliant family. Typically, there are only two circuits. The combined amperage on the fuses must not exceed 30 amps, resulting in combinations of one 10-amp circuit and one 20-amp circuit, or two 15-amp circuits. This amperage is sufficient for a refrigerator, television, stereo, hot water heater, and other basic necessities; but it will *not* accommodate several small appliances turned on simultaneously, or even one heavy-duty appliance. This can really pose some problems if you decide to extensively renovate your home on the existing service.

Somewhat newer homes have three-wire, 120-240 volt, 30-amp service. Three-wire service more than doubles the electricity available from a two-wire system. It permits at least twice the number of 120-volt circuits and an additional 240-volt circuit. The increased capacity accommodates the use of more small appliances simultaneously, as in a kitchen busy at breakfast. However, the low amperage level precludes the use of a major appliance like an electric dryer, which alone requires 40 amps, despite the increased voltage offered by the 240-volt circuit.

An even more sophisticated service is that of the three-wire, 120-240 volt, 70-amp accommodation. There are enough circuits for all those electrical wedding presents, and there is also enough voltage and amperage to support an electric dryer and electric range.

Improving the electrical service is one of the most effective ways to modernize an old house without harming the architecture. It is generally advisable to upgrade 30-amp service, whether two or three wire, to 100-amp service. Generally, it is unnecessary to upgrade a 70-amp system that is otherwise operating safely. The going rate presently charged by a licensed union electrician for increasing service from 30 amps to 100 amps is $300 to $375. This includes the installation of a new circuit breaker panel and all the wiring upstream (on the service side) of that point. The additional wiring (on the house side) that makes the upgraded service utilitarian costs roughly an additional $25 per circuit and $39 per socket.

Two-wire service, and the original two-wire portion of an upgraded three-wire system, were typically installed by the knob-and-tube method. A conductor wire wends its way through the house, with a porcelain knob to support the wire where it crosses wood. A porcelain tube encases the wire where it crosses other wire and in vertical runs through wood members.

In modern or modernized residential wiring systems, BX and Romex wire have replaced the knob-and-tube system. BX wire encases two wires in an armor shield, and Romex wire encases two wires in a nonmetallic sheathing. Although BX and Romex are more up to date, knob and tube, when properly installed, is a safe and efficient system that does not require repair. However, when extensive problems exist in a knob-and-tube system, it is practically impossible to repair it in kind because the new or good-as-new parts required by the electrical code are difficult to get.

NEW INSTALLATION

All new electrical installations must conform with the National Electric Code and special local, town, or county requirements. Existing installations are acceptable if they have been safely maintained and the original integrity of the system has not been altered or abused. For example, overloading is one type of abuse that could have damaged the electrical system, whether or not alterations were involved.

Whereas potential dangers from carpentry and construction work can usually be foreshadowed by warning signs (the joist begins to crack, the ladder to rock, the board to slip, for example), they are not all that obvious when working with electricity, because current cannot be seen as can a crack in a board or felt as when a board weakens. Construction accidents can be prevented and often rectified once they do occur, but working with electricity is not as manageable. If you make the slightest error, the mistake can be tragic or fatal. Electric shock and fires caused by lack of electrical knowledge can strike without a moment's notice and take a severe toll on both property and life.

The best insurance against personal injury or property damage is to hire an electrical contractor. When seeking one, keep in mind these rules:

1. Always look for a State Electrical Contractor's License, C-10.
2. Always get competitive bids in writing.
3. Never pay more than 10 percent of the fee in advance. You are entitled to this limitation by the State Contractor's Law.
4. Never make the closing payment until final inspection is made by the city or county and the work approved.

In the state of California, if you sell the house within a year from the time you bought it, you are required to use a licensed contractor for the electrical work. That way the buyer has recourse should anything go wrong. If you are uncertain about your electrical system and its condition, many cities or counties will inspect it and provide a written report of any problems, inadequacies, code violations, along with recommendations for certain changes. The fee is nominal and well worth an hour of your time.

(WIRING)

When the level of your electrical service is upgraded, it is time to add new circuits. When the present service is adequate but inconvenient and underutilized, it is time to add a new circuit, or at least a more convenient socket, switch, or fixture to an available circuit. When a portion of an otherwise acceptable electrical system breaks down, it is time for repairs. Each situation requires work behind the walls of an existing house, so wiring becomes a matter of carpentry as well as electrical know-how.

The technique used to wire or rewire behind solid walls can have little effect on the architecture or can totally devastate it. It is up to the homeowner to protect the house from the electrician who pokes holes, slashes molding, or recommends ripping out an entire wall. There is no need to destroy good lath and plaster to rewire, nor to gut the innards of a house to modernize an electrical system.

GUIDELINES

Remember, the more walls that are ruined to make the electrician's job easier, the more difficult or expensive it will be for you to replace or repair them. To keep the cost of wall repairs from inflating the cost of wiring, and to safeguard the architecture of the house, follow these guidelines:

1. Lay out the route of the wiring yourself, before the electrician goes to work. Compare notes.
2. If one or more walls are slated for repair, take advantage of the exposed frame to accomplish the necessary electrical work in that area.
3. Whenever possible, route the wires through hidden passages (crawl space, pipe chase, abandoned vents) instead of through the wall itself.
4. For ceiling runs, like wiring an overhead light fixture with a pull chain into a switch, send the wire parallel to the joists in the void between them. This saves you poking and drilling to find the joists.

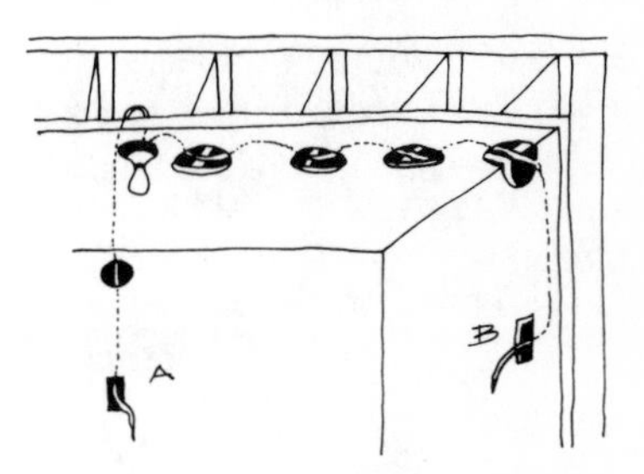

A
B
SWITCH LOCATION "A"
THIS IS THE EASIEST. WIRE IS RUN IN THE CAVITY BETWEEN JOISTS. ONE HOLE IS MADE TO FISH THE WIRE THROUGH.
SWITCH LOCATION "B"
AVOID IF POSSIBLE. A HOLE MUST BE CUT IN THE CEILING AT EACH JOIST TO ALLOW NOTCHING FOR WIRING CLEARANCE.
NOTE: INSTALL PROPER SIZED ELECTRICAL BOXES AT SWITCH & FIXTURES.

Adding a wall switch for an overhead light

5. For bottom-of-the-wall runs (for example, a series of convenience sockets) remove the baseboard and scrape a channel in the plaster behind it as a bed for the wire. If the wire is closer than 1 inch to the surface, code requires that the wire be covered with a metal plate to protect the wire from piercing nails. Replace the baseboard after the wire is inspected. Elaborate cornice moldings can also hide wiring.
6. For first-floor sockets, use the open cellar ceiling for the wire, and come up into the room behind the baseboard.
7. For vertical-wall runs that require notches in obstacles

like studs and firestops, use adjacent rooms or closets where patched surfaces are less noticeable, less important, or will be covered by wallpaper.

8. Go fishing. *Fishing* is the electrician's art of sending wire through one end of a wall or ceiling and mysteriously retrieving it out another. Fish wire is steel tape 3/16 inch wide and 1/16 inch thick. It is stiff enough to maintain direction when being forced through an invisible passage, yet flexible enough to bend around corners. As with the sport, the essential skill in fishing is patience, and there is no need to pay an electrician's rates for that.

9. Resort to surface wiring only where exposed circuits will not detract from the architecture or will give the room a jury-rigged appearance. For example, in a kitchen, multiple outlet strips along countertops can be very convenient. They are also in keeping with the appliances they serve, and their obtrusive appearance can be acceptable. However, in most other rooms surface wiring along the wall or baseboard looks amateurish.

10. Locate switchplates, sockets, and other surface hardware so it is entirely in the wall or baseboard, but not half and half. Never place a switch or socket in the door trim.

[SWITCHES AND PLUGS]

You may discover that your wiring need not be tampered with; only a few wall sockets and switches are faulty. They either fail to operate, or they operate inconsistently. If this is the case, replace the switches and plugs. This is one area of electrical work that really can be done by the amateur. However, do be careful, for even if the job is rather simple, you can still get an electric shock or start a fire if you botch the job.

REPLACING A SWITCH

Before attempting to probe behind the switchplate, first be sure to *deactivate* the fuse or circuit breaker of the circuit that controls the switch. (Turn off the breaker or remove the fuse.) Then remove the screws from the face of the switchplate. The plate will come off. Now remove the mounting screws that keep the switching mechanism in position. These are at the upper and lower ends of the switch. The switch will "pop out" of the box a little, but pull it out far enough so that you can disconnect the wires. Carefully note which wires connect to what; in fact, label the wires or draw a sketch so you can keep track of the wires.

After you remove the faulty switching mechanism, replace it with a similar switch. If in doubt as to what is similar, ask for assistance at the hardware store when

Replacing a wall plug

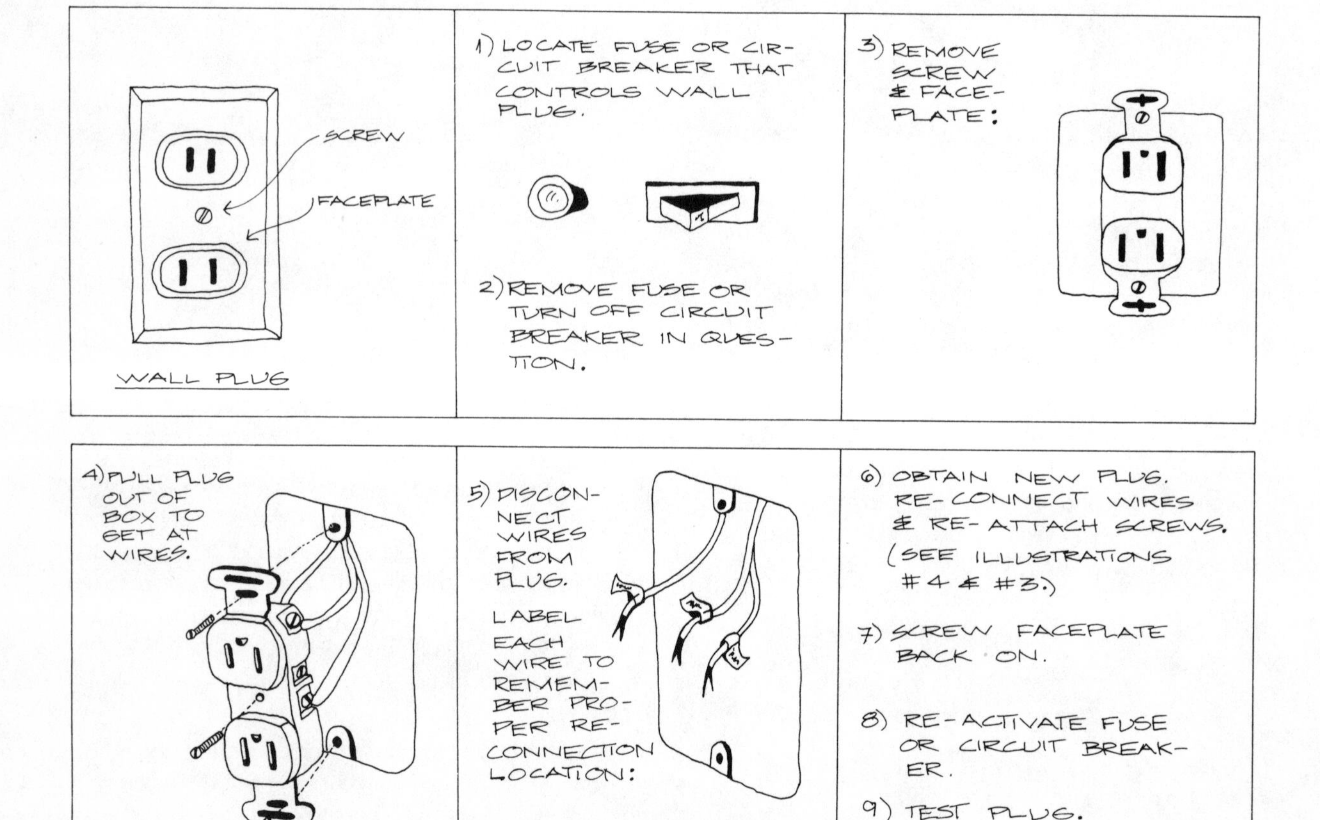

you purchase the new element. (Take the old one with you.) To install the new switch, reconnect the correct wires to the proper screws. (You will know where to place them if you properly noted their locations before you disconnected them.) Tighten the screws enough so that the wires do not come off. Then, just reverse the process by reinstalling the mounting screws and those that fasten the switchplate. Now reactivate the breaker or fuse. Turn the switch on and off to test it; it should work properly.

Wall outlets work in just the same manner. Follow the instructions for replacing the switch, and above all *take the necessary precaution of deactivating the correct circuit first!*

[LIGHT FIXTURES]

An electric light fixture is essentially a decorative shell for wires, socket, and bulb. Wherever possible, you should retain and use the house's light fixture. If the illumination level is insufficient, try to supplement the original light with extra fixtures elsewhere in the room rather than substituting a new fixture for the old one.

Prior to the introduction of electricity, gas was used to power the light fixtures. As a result, the fixtures needed to be rather heavy and massive. They were made of cast iron and white metal, with ceramic ornaments and etched shades. Later Victorian and some Colonial Revival houses built or "modernized" subsequent to electrical service had fixtures with separate electric lights and gas jets, just in case. The fixtures were made of brass or

Replacing a wall switch

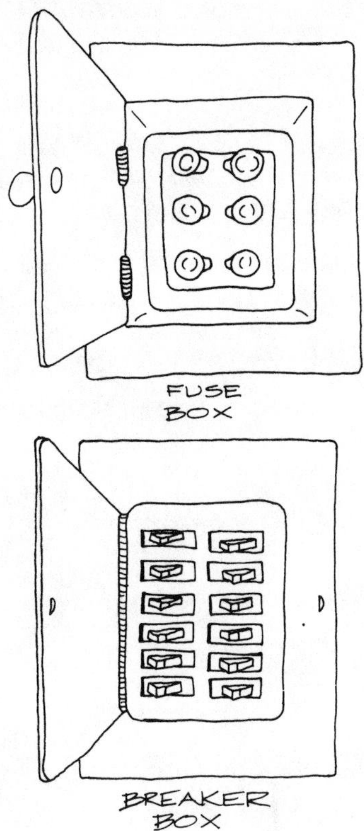

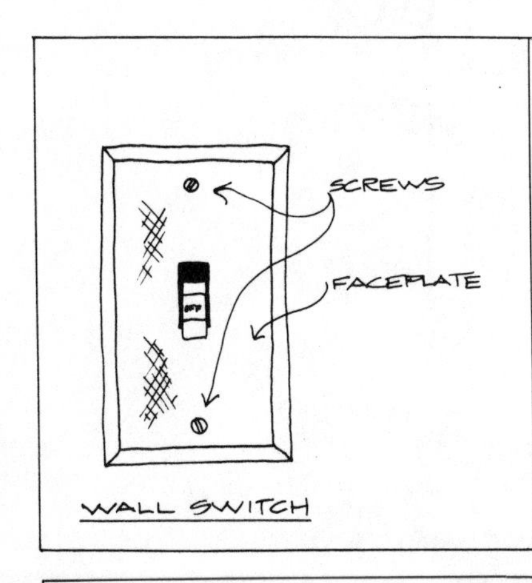

1) LOCATE FUSE OR CIRCUIT BREAKER THAT CONTROLS WALL SWITCH.

2) REMOVE FUSE OR TURN OFF CIRCUIT BREAKER IN QUESTION.

3) REMOVE SCREWS & FACEPLATE:

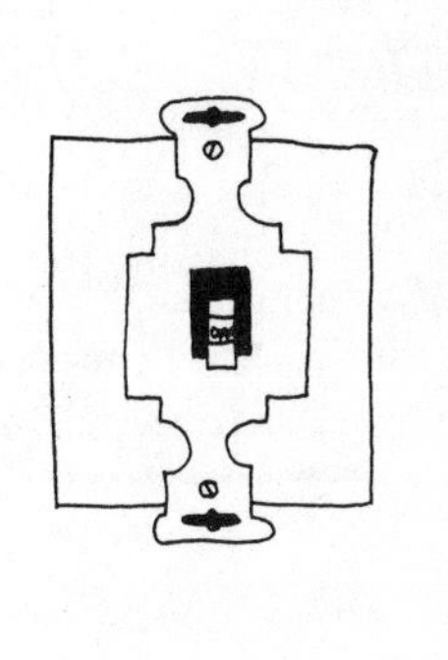

4) PULL SWITCH OUT OF BOX TO GET AT WIRES:

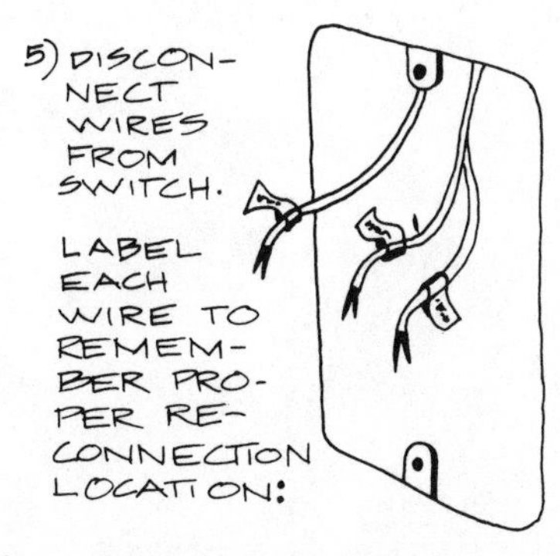

LABEL EACH WIRE TO REMEMBER PROPER RECONNECTION LOCATION:

6) OBTAIN NEW SWITCH. RE-CONNECT WIRES & RE-ATTACH SCREWS. (SEE ILLUSTRATIONS #4 & #3.)

7) SCREW FACEPLATE BACK ON.

8) RE-ACTIVATE FUSE OR CIRCUIT BREAKER.

9) TEST SWITCH.

Socket and switch: do's and dont's

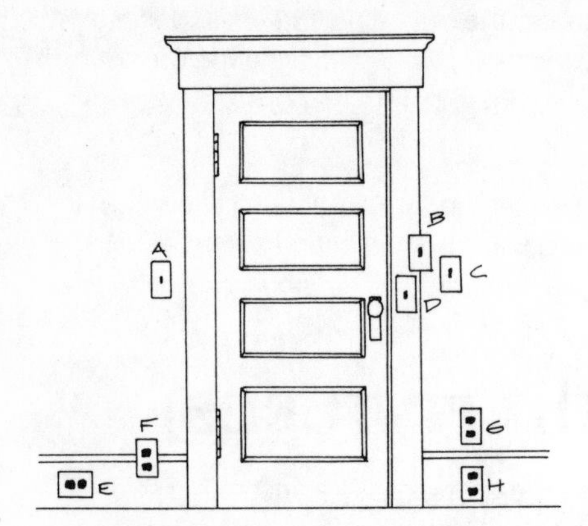

SOCKETS E & H ARE PREFERRED (CORDS ARE UNOBTRUSIVE, FACEPLATES ARE CONTAINED IN BASEBOARD.) G IS ACCEPTABLE. F DEFACES MOULDING.

SWITCHES – C IS THE BEST LOCATION. B & D MAR THE DOOR TRIM; A IS ON THE WRONG SIDE OF THE DOOR SWING.

FISHING PREVENTS UNDUE DAMAGE TO WALLS, CEILINGS & ORNAMENTATION WHEN THERE IS NO EASY WIRING ACCESS. THE FOLLOWING SKETCHES SHOW THE BASIC TECHNIQUE. SOLID BLOCKING ABOVE THE WALL, FIRESTOPS BETWEEN STUDS & LUMPS OF PLASTER ARE AMONG THE HIDDEN OBSTACLES THAT MAY COMPLICATE AN ALREADY DIFFICULT PROCESS. PARTICULAR PROBLEMS & VARIANTS ON THE TECHNIQUE ARE ALSO LIKELY.

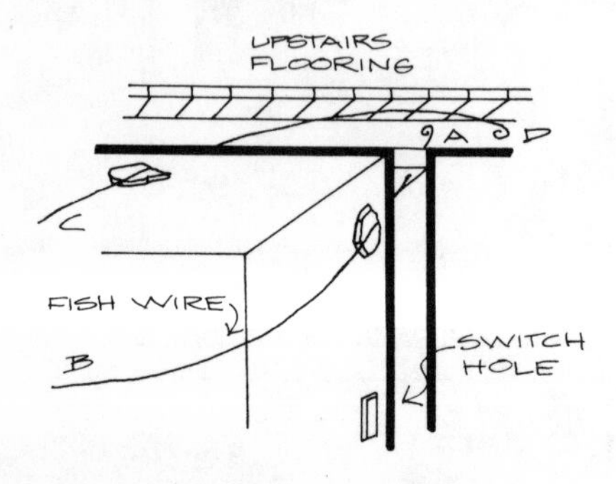

• OPEN A HOLE IN THE WALL 5" OR 6" FROM THE CEILING. WITH AN 18" DRILL BIT, MAKE AN ANGLED HOLE INTO THE CAVITY ABOVE THE CEILING. RUN FISH WIRE INTO THE CAVITY FROM THE CEILING FIXTURE HOLE AND THE WALL HOLE. "FISH" UNTIL HOOKED ENDS "A" & "D" INTERLOCK (PERSISTENCE & PATIENCE ARE NEEDED HERE). PULL ON END "C" UNTIL "A" EMERGES FROM CEILING HOLE. ATTACH ELECTRICAL CABLE TO HOOK "A", MAKING SURE THE CONNECTION IS NOT TOO BULKY TO PASS THROUGH THE DRILLED HOLE.

bronze, and because the metal was shaped, rather than cast, the form of the fixtures was more graceful. The shade was opalescent or of clear glass.

After the turn of the century, electrical lighting became increasingly common, and by around 1915 it was looked upon as far less than a novelty. Initially, the electric fixtures looked a lot like the earlier gas fixtures. Often they were made of square brass stock, and the oxidized finish was introduced. For shades, artglass, such as the Tiffany dome, became popular, although the utilitarian bare bulb persisted into the 1920s.

If you have original fixtures or remnants, you can have them made serviceable by just installing new, inexpensive electrical parts. If you are a bit hesitant about making the transformation yourself, an electric repair shop can do the work for you, usually for less than $10.

PULL ON END "B" UNTIL THE CABLE APPEARS AT THE WALL HOLE. RUN FISH WIRE UP FROM SWITCH HOLE TO WALL HOLE. FASTEN CABLE TO FISH WIRE & PULL DOWN TO SWITCH HOLE.

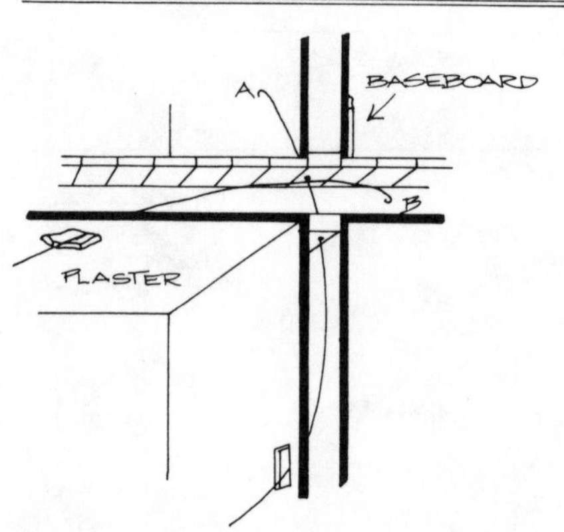

• WHEN THERE IS A WALL DIRECTLY ABOVE THE SWITCHPLATE WALL, THE EXTRA WALL HOLE IS UNNECESSARY. REMOVE THE UPSTAIRS BASEBOARD & DRILL AN ANGLED HOLE INTO THE LOWER WALL CAVITY. FEED FISHWIRE INTO WALL & OUT SWITCH HOLE. PULL END "A" BELOW UPSTAIRS FLOORING. LOCK ENDS "A" & "B" AND PROCEED AS IN THE FIRST EXAMPLE, BUT DRAWING THE CABLE DIRECTLY OUT OF THE SWITCH HOLE.

REPLACEMENT GUIDELINES

When seeking replacement fixtures, you may run into a bit of a snag with the law. Many historical fixtures do not meet current electric code qualifications, and are not Underwriters' Laboratories (UL) listed, as so many ordinances require. The best compromise is to find a reproduction of the original, one that resembles the historically correct fixture but also meets today's safety standards. Some reproductions look phony and really can detract from other carefully labored areas. However, other reproductions, for a bit higher price, do not look gaudy at all.

It is well worth the extra silver for a look of gold as opposed to plastic. Do not resort to moon globes or other smooth or flush types of lighting, which are reminiscent of condominium kitchens and matchbox-apartment bathrooms. You may be tempted to consider them because they cost less than more desirable fixtures. Or you may feel that rather than choose a costly fixture which is an obvious fake, why not "fake all the way" since the contemporary design is simple, subtle, and does not draw attention to itself. This contention is wrong. A light fixture, when lit, *does* draw attention to itself. It is the contrasting element in the environment. The obviously foreign form of a contemporary fixture *will* make a point, much more so than any imitation antique.

When modern fixtures are necessary to replace or supplement originals, follow these guidelines:

1. Determine the function of the light. Is it for general illumination? Is it light to work by? Is it concealed or indirect light? Is it a spotlight? Is it for security?

2. Determine the quality of light desired: fluorescent or incandescent. Incandescent light is warmer than fluorescent, and also seems more natural. Unlike fluorescent light, incandescent light throws a shadow, because the light is directed. As a general rule, use incandescent light indoors. Reserve fluorescent light for work areas and for indirect lighting, for instance, behind a valance. In a kitchen, fluorescent light makes food look unappetizing. Wherever fluorescent lighting is a must, use the warm white type to minimize its inherent disadvantages.

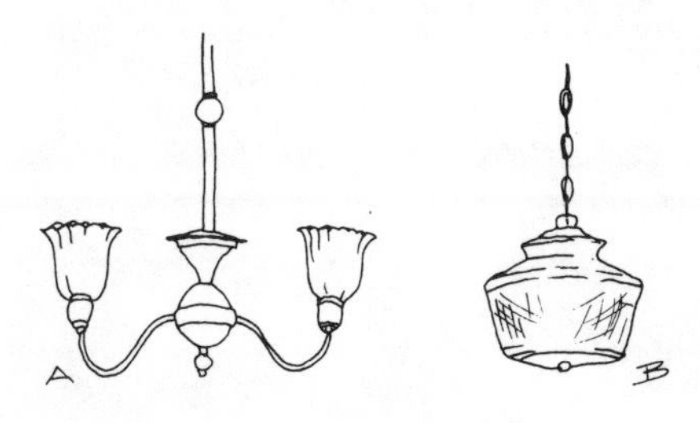

Suitable light fixtures

Unsuitable light fixtures

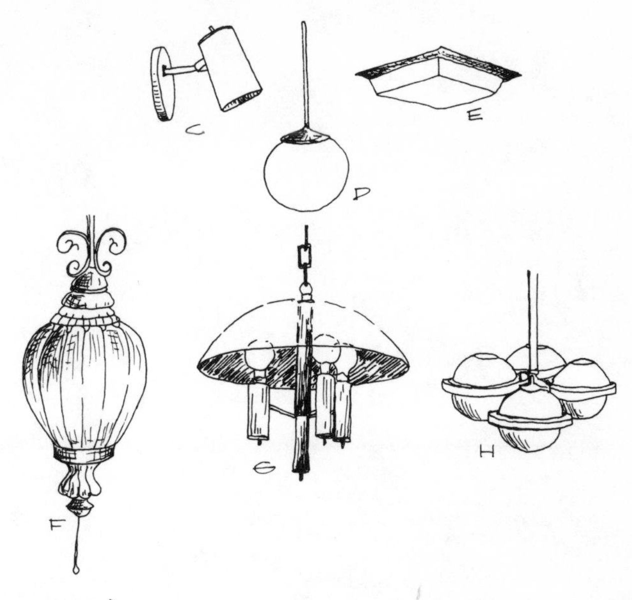

3. Select a fixture in scale with the room and the wall. If the ceiling is high, use a pipe or pole fixture to bring the chandelier down to a practical level. If the room is small, use a wall fixture that minimizes its intrusion into the space. Do not use a gigantic or ostentatious lamp in a small room or it will be the focal point.

4. Install the fixture high enough to cast light down on the people and objects in the room. Immobile wall fixtures should be above eye level but below the cornice molding.

5. Do not feel obligated by convention to install a central ceiling light. Although it provides the convenience of illuminating a dark room by a switch at the door, it is otherwise relatively useless in providing good light for specific activities. Besides, other lights can be rigged to turn on at the flick of the door switch. A combination of wall lights and floor or table lamps much more effectively illuminates work or leisure areas. If there is a ceiling socket and it is time for a new fixture, consider a fixture you can equip with a dimmer switch, which enables you to create a different mood for different occasions.

(CHAPTER 15)

PLUMBING AND HEATING SYSTEMS

The plumbing system is made up of two separate lines. The fresh water supply line brings water into the house under 50 to 80 pounds per square inch of pressure from the main in the street. The supply line is divided at the hot water heater. Part of the incoming flow is heated there, and from there pipes with hot and cold fresh water run parallel to one another and serve the plumbing fixtures throughout the house. The second line is the drain-waste-vent line (DWV), which carries used water out to the sewer by gravity and noxious fumes up into the air at roof level.

(PIPING)

In most pre-World War II homes, the water supply line is constructed of galvanized steel, and the DWV line of cast iron. Galvanized pipes may become fragile with age, due to corrosion and frozen fittings, but if the plumbing system is in good or repairable condition, there is no need to update it to copper or plastic. Additions to the existing system can be made as long as a proper transition fitting is used.

The service from the water main to prewar houses is typically a ½-inch pipe. In 1949, ¾-inch pipe became the standard. When the service pipe becomes encrusted, the inside diameter is considerably smaller than its name implies. It is necessary to consider upgrading plumbing service from ½- to ¾-inch pipe only if you experience a disturbing reduction in the quantity of water available when two fixtures are used at once, or if you are building an additional bathroom.

The connection between the DWV line and the sewer on the street has a tendency to break, and yours probably *is* broken if you happen to be in the one out of every four households that this occurrence afflicts. To detect a broken connection, look for signs where water might be unnaturally nourishing the surrounding area; this is where the leak is. Broken, moist concrete, sudden sprouts of greenery in an otherwise barren spot, or rodent burrows are several indications. If you have a broken sewer line, call the Department of Public Works so they can visit your sewer and determine the nature and extent of the problem. Do not let this problem slide by, because it can only worsen with time. Improper disposal of waste means an unsanitary environment; and if your home is in an urban area, you know that rampant filth and uncontrolled waste disposal can quickly lead to an infestation of rats.

INSTALLATION AND REPAIR

Most of the plumbing system is hidden and so has a limited bearing on the architectural character of the house. However, as with the electrical system, the installation and repair of pipes in an older house is a matter of carpentry as well as plumbing know-how. A drill with a hole saw bit, for example, is as handy as a pipe wrench.

When adding new pipes or when improving the old ones, do as little damage as possible to the original materials and spaces of the house:

1. Installation of a second-story toilet or bathtub requires a P-shaped trap (called a P trap) below the floor. Conceal this pipe between the joists; never lower the first story ceiling to hide it.

2. If possible, use existing holes in the wall or floor for pipe connections; or if making new holes, install them out of sight behind the cabinets.

3. Use hidden passages, crawl space, abandoned vents, and the like to hide new runs of pipe and minimize the destruction of the valuable wainscoting or plaster walls.

4. On the exterior, locate pipes and vents as inconspicuously as possible, preferably at the rear or on hidden sides of the house. Paint the pipe the same color as the wall behind it.

You may notice a preexisting network of old cast iron drains on the outside of the house. This is not a true historical design feature. Many old homes were erected before modern plumbing. When plumbing was installed, the easiest solution was to add much of the system onto the exterior, in the same fashion as wires on the outside of walls.

(FIXTURES)

The most obvious parts of the plumbing system are the fixtures in the kitchen and bathroom. These two rooms, in that order, are the most commonly remodeled rooms in the American house. As a result, valuable porcelain and vitreous china sinks, tubs, and toilets, many with luxurious brass hardware, were removed in favor of something more modern although less sturdy. If the original plumbing fixtures are still in place, do retain them. For example, one of the most exciting and unique bathroom fixtures is a pull-chain toilet. If it works properly, it is an asset to the authentic antiquity of your home.

A plumbing permit is required for all plumbing other than minor repairs. Except for complicated mid-run piping, plumbing is within reach of the technically adept homeowner. Approximately one out of four plumbing permits is issued to the do-it-yourselfer.

Some jobs are beyond the lay person's ability. When they are, hire a plumber, one who has a C-36 Plumbing Contractor's plumbing permit and one who will do the work.

LEAKS

Basic repairs of aging faucets and other fixtures require only some knowledge of the mechanisms. One problem often plaguing older homes is a water leak. If you suspect one, attempt to find out if it exists by first turning off all the faucets and water outlets in the home, including outdoor outlets. Then go to your water meter and observe the pointer and the readings for about 15 minutes to 25 minutes. If the pointer moves the least bit, you know there is a leak in one or more places. If you determine that there are no visible leaks, your problem requires a professional's probing and repair. Chances are

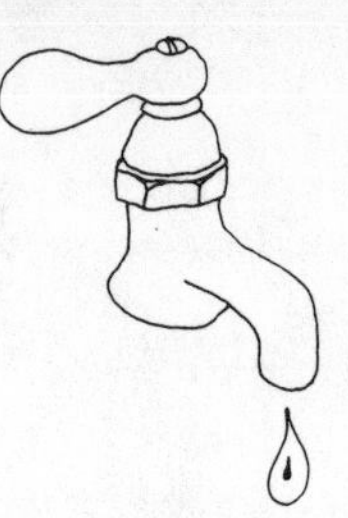

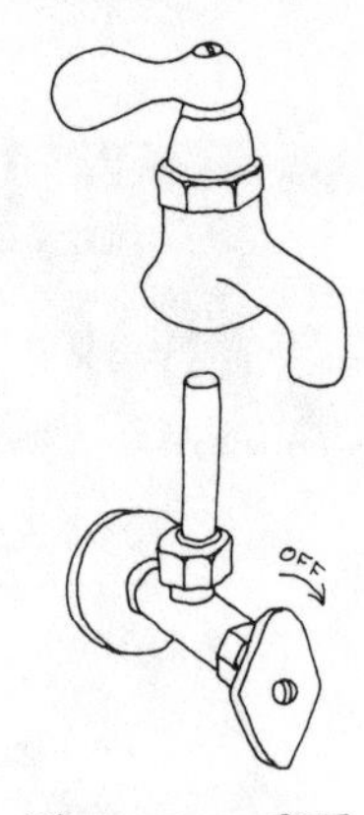

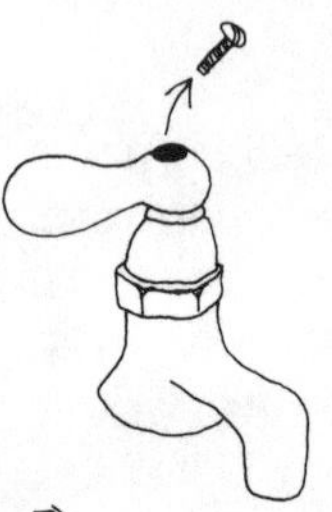

1) FAUCET

2) TURN OFF SHUT-OFF VALVE BELOW SINK.

3) REMOVE SCREW FROM TOP OF FAUCET & TAKE OFF HANDLE.

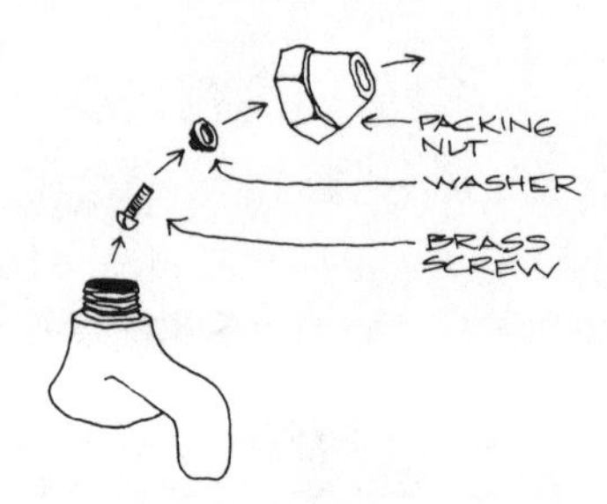

PACKING NUT
WASHER
BRASS SCREW
4) REMOVE PACKING NUT, WASHER & BRASS SCREW.
5) INSTALL NEW WASHER & BRASS SCREW.
6) RE-ASSEMBLE FAUCET BY TIGHTENING DOWN PACKING NUT & TOP SCREW.
7) TURN ON SHUT-OFF VALVE BELOW SINK.
8) TEST FAUCET.

Fixing a dripping faucet

the leak is in such a place that it could ultimately damage some portion of the house.

DRIPS

Drippy faucets are the most common and annoying problem. If your spout is leaking, first shut off the valves located below the sink. (If both faucets are leaking, turn off both valves. If only one faucet is leaking, shut off only the related valve.) Then remove the cap and screw from the faucet, and with an adjustable wrench take off the packing nut. Now remove the brass screw and washer. The drip problem is usually caused by a worn washer and brass screw. Replace these elements in kind and reassemble the faucet. As with rewiring a switch, note how you undid the mechanism so putting it back together will not be a puzzle.

PORCELAIN DAMAGE

If the porcelain finish on the toilet, sinks, and tub is wearing thin, have them refinished. This is not too expensive (beginning about $85), and it is worth the money in relation to the value it gives your home. But before deciding to resurface the fixture, examine it closely; it may need just a good cleaning. Reduce stains with a Clorox solution. Use about ¼ cup to 1 gallon of water, and let the solution soak the stains for 24 hours. You can correct the cracks and chips with a porcelain patch. This is usually done by a company specializing in surfacing damaged areas. They use a high-gloss plastic that rejuvenates the entire fixture's appearance.

A temporary improvement to dings in porcelain is a special epoxy glue, but those who have tried it report mixed results. Success depends a great deal upon the efficiency of cleaning the surface first. The easiest option is to learn to love the cracks as badges of antiquity. The best care for porcelain is to keep it clean with a nonabrasive cleanser.

CROSS CORRECTION

An older sink, toilet, or tub may have a cross connection, and this demands correction. A cross connection occurs when the supply line is placed in so low a position on the fixture that accumulated used water could enter the fresh water line and contaminate it. For a toilet with a cross-connection, install a new Fluidmaster ballcock, which costs about $7. If the faucet of the tub is below the rolled tub rim, you can replace the spigot with a spout that arches below the water line, although swan-neck spigots are fragile. Check codes, because in many cases the spout must be at least 1 inch above the overflow.

CLAWFOOT TUB

You can convert clawfoot tubs into a combination shower and bath without modifying the pipes. A single piece of hardware, with faucets, extension pipe, and showerhead is available at plumbing-supply stores. It costs from $35 to $65, depending upon the brand. These stores usually also carry circular shower curtain rods.

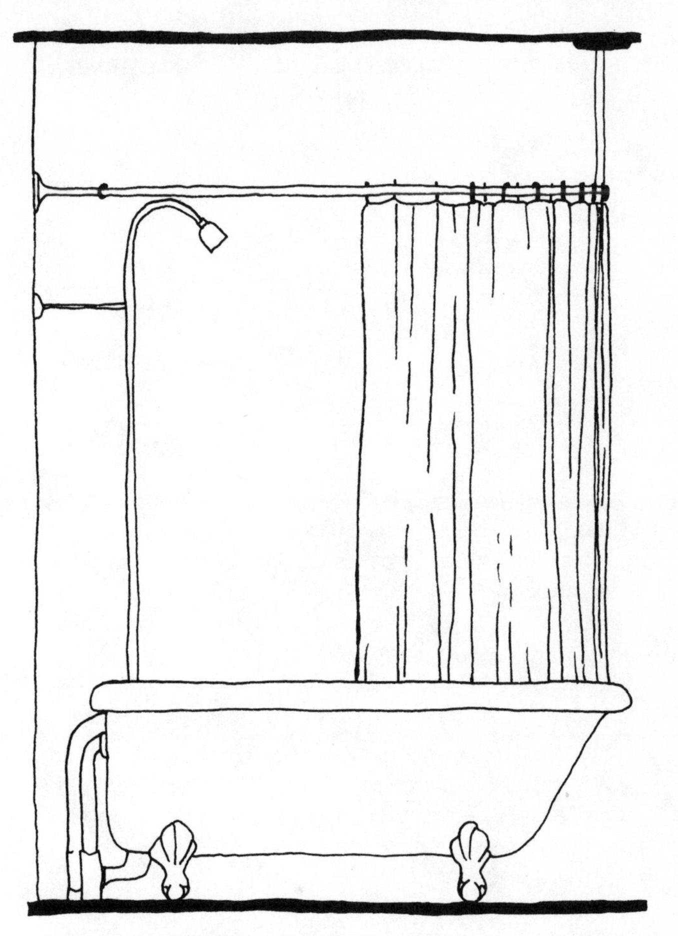

Clawfoot tub with added shower

If the clawfeet are covered by so many layers of paint that they look as if they are wearing socks, use a chemical stripper to disclose the detail. Mask the area around the tub carefully. Use an enamel spray paint suitable for metal to repaint the feet and the outside of the tub. If masking is not possible, use a brush-on enamel paint suitable for metal, like Rustoleum, and a good-quality brush with fine bristles; a cheap brush will leave stroke marks.

It is preferable to keep the clawfoot tub unenclosed and to display it like a piece of sculpture, but some people choose to box it in. If you do, be certain to leave a 12 x 12-inch hinged access panel to the above-floor pipes at the head of the tub, because these pipes require occasional attention.

REPRODUCTIONS

If the original bathroom accessories are gone, simplicity is the rule when selecting modern fixtures. If you cannot get at least a reasonable facsimile, by no means substitute some antagonistic style just because it looks antique. The market is full of tacky imitations, so beware. Look for *authentic* materials; marbleized plastic that imitates stone is not only a hearty fake, it costs too much.

To replace missing porcelain fixtures with the real thing, try salvage yards or antique shops. Also, expensive but good reproductions are available at specific suppliers.

In the kitchen and laundry area, a common problem is dry rot in the wood structure and around sinks. If the original design did not include an effective splash area, the wall and floor can become deteriorated from continual exposure to water. Correct dry rot around sinks before replacing the sinks, and be sure there is adequate drainage and waterproofing to prevent reoccurrence of the same situation.

[HEATING SYSTEM]

The mechanical (heating) system is so named because it involves machines, unlike electricity and plumbing, which are essentially delivery systems. In the home, the mechanical system manages the distribution of air: heating, ventilation, and air conditioning. (In the industry, this is commonly referred to as HVAC.) With the few exceptions noted below, improvements to the mechanical system have limited bearing on the architectural character of the house, so the repair aspects are not addressed in detail here. If the mechanism is faulty, parts can be replaced, or a whole new system can be installed.

FLOOR FURNACE; WALL HEATER

The floor furnace is associated with the Classic and California Bungalows, but it was installed in new houses well into the 1950s. There is nothing intrinsically wrong with a floor furnace, but the Federal Housing Authority (FHA) frowns upon them, as do other experts who feel the location is inappropriate for safety. They are an obstacle, especially for young children, who can fall and get burned, or for anyone walking in the dark and unaware of the heater's presence. If you have such a furnace, check underneath the house to be sure that the pipe which vents the furnace is still connected and in good repair. These furnaces are prone to rusting, and unless their condition is periodically examined, a malfunction can go unnoticed. At the insistence of FHA, floor furnaces are often replaced with vertical wall heaters, which are safer but infringe upon the original architectural statement. The best solution, if any, is to direct visual attention from the wall heater to other areas within the vicinity.

RADIATORS

Steam radiators in older houses can be a symphony of noises. Replacement of the central furnace will do nothing to stifle the hisses, gurgles, and thuds because they are generated by the steam-distribution mechanisms: the pipes, radiators, and air valves.

EXPANSION OF SYSTEM

Additions to an inadequate heating system invariably have had a visual impact, so try to minimize the intrusion. Consider upgrading the system you already have by installing supplementary registers, for example, rather than switching to a brand new setup that engenders exposed duct work or bulky units. If you must introduce a new permanent heating device to an unserved room, remember that baseboard heaters are the least obtrusive. The Intertherm type is considered especially good by some experts because of its safety, control, and humidity features. Note that a mechanical permit is required for all such work (other than minor repairs).

Before expanding the heating system in the house, be sure that the building is properly insulated to prevent heat loss. You may find that insulating your home can make such a substantial difference in the amount of heat you are able to retain (or repel in the summer), that a new or expanded HVAC system is really not warranted.

For safety purposes, make sure all gas appliances are vented to approved capped flues that terminate above the roof. As evidence of an appliance malfunction, look for sooting atop the vent. Any other doubts you have can be cleared up by a survey and analysis by the city's or town's Inspectional Services Department.

(CHAPTER 16)

THE FIREPLACE

Throughout history, the hearth has been a romantic symbol of a happy home, a warm symbol that beckons the world weary. The Victorian fireplace provides both sensual enjoyment and visual focus. An attractive and functional fireplace definitely add resale value to a house—some realtors estimate as much as $5,000. Do not board up an old fireplace in an attempt to modernize; to do so drastically alters the character of a room, needlessly sacrifices an economic asset, and deprives the room of the special character only the fireplace can offer. Besides, in these days of energy conservation, why not take advantage of an auxiliary source of heat?

An important note: Before you buy a house, evaluate the condition of the fireplace and chimney and estimate repair costs; use this as a consideration when negotiating the sales price. The cost of making a fireplace operative can really be a factor in agreeing upon a deal, as much as the cost of termite damage and repair.

(STYLE)

The style of the fireplace is dictated by the architecture of the house. It could even be said that the fireplace sums it up because the features typical of the building design—proportion, mass, materials, and ornamentation—are incorporated into the mantel, firebox, and hearth. If the original mantel, surround, firebox, or hearth is missing or irreparably damaged, replace the missing or damaged part in a manner sympathetic to the character of the house.

Basically, there are two kinds of fireplace designs: the built-in mantel and the add-on mantel. The latter is typical to Victorian houses. During construction, a hole was roughed into the wall; a firebox, damper, and flue were installed; and the surround was surfaced with fire-resistant tile. Only much later was the mantel added. Thus, the Victorians could select whatever mantel design appealed to them, as if they were choosing a piece of furniture. The distinct disadvantage of the add-on mantel was that it facilitated removal of mantels in subsequent years by misguided modernizers and by well-informed vandals, who sold them.

Mantel designs were quite varied, based on the assortment of construction materials, mail-order catalogs, and Victorian tastes. The earliest examples were carved from marble, but this costly stone was soon replaced with its economic counterpart, slate. By the 1880s, carbon copies of marble mantels were hewed from wood. In San Francisco Stick houses, an elaborate overmantel, with mirrors and whatnot shelves, was incorporated into the fireplace design.

Most Victorian fireplaces were the primary heat source and generally burned coal. With the introduction of central heating, fireplaces were redesigned as an auxiliary heat source. They were also made larger to accommodate wood as a fuel, and the mantels became built in. In Colonial Revival houses, both types of mantels were used. The predominant mantel design is a pair of classic columns with a pediment on top, in darkly stained or painted wood.

The mantel was usually flanked by built-in bookcases with glass doors, and the stack by a pair of fixed windows, often leaded or stained. The hearth was made of brick or square tiles in gray, tan, or brown. The mantel was built entirely of clinker brick or of wood, with the hearth tiles incorporated as the surround. To determine whether the wood was originally painted, look at the abutting cabinet shelves. If their undersides were painted, the mantel usually was also, and the wood probably does not warrant stripping.

(REPAIR)

To replace a missing add-on mantel, scavenge salvage yards and antique shops. Because of the Victorian eclecticism regarding mantel selection, you need not find

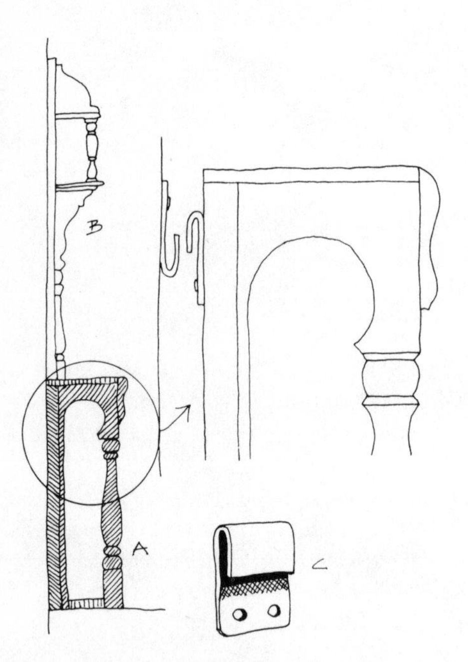

B
A
C
ADD-ON MANTELS (A) & OVERMANTELS (B) CAN BE SECURED TO THE WALL WITH FLAT HOOKS (C). USE 2 OR MORE SETS OF HOOKS. PATIENCE & CAREFUL MEASURING ARE NEEDED TO ENSURE THAT THE MANTEL WILL JUST REACH THE FLOOR.

Add-on mantel

one that is a precise microstructure of your home's architectural style. Just make sure that it is *reflective* of your home's architectural style in its detail, proportion, or scale.

To reconstruct a damaged built-in mantel, hire a cabinet maker, or do it yourself. Build a simple wood box with a ledge on top, perhaps with chamfered corners. For ornamentation, add standard molding, separately or in creative combinations (as explained in the Trim section in Chapter 12). To incorporate hearth tile in the mantel ornamentation, cement it in place with a product suitable for wall tiles, and frame it with moldings.

Reproduction of the original mantel is not essential for a well-designed result; a reasonable facsimile can allow

Rescuing a remodeled mantel

IN THIS MISGUIDED REMODELING, IMITATION BRICK HAS BEEN APPLIED, A SKIMPY MANTEL REPLACES THE ORIGINAL, & THE REMAINING PANELING HAS BEEN PAINTED & THE CORNICE REMOVED. OTHER TECHNIQUES ARE USED TO ALTER FIREPLACES, BUT LOSS OF CHARACTER OFTEN RESULTS.

RESCUING THE FIREPLACE FROM THIS SADLY COMMON DISGUISE ENTAILS REMOVING THE BOGUS MATERIALS & ADDING MANTEL, BASEBOARD, & CORNICE MOLDING THAT MATCHES THE SURVIVING CRAFTSMAN PANELING IN PROPORTION. (STRIP THE PANELING FIRST.) UNGLAZED TILE IS USED FOR HEARTH & SURROUND.

you to stray a bit from strict adherence to present-day code. In contrast, a redesigned and rebuilt mantel must conform to all the standards of today's building codes. However, these requirements do not preclude replacement sensitive to two historical factors:

1. Clearance of at least 12 inches between the fireplace opening and any combustible materials. (In other words, a 12-inch-wide facing of fire-resistant brick or tile between the fireplace opening and the wood mantel.)
2. Hearth dimension no less than 20 inches in front and 12 inches to either side of a fireplace opening that measures 6 square feet or more; hearth dimension no less than 16 inches in front and 8 inches to either side of a fireplace opening less than 6 square feet.

Whether replacing or rebuilding a mantel, follow these guidelines:

1. The size of the mantel should be in proportion to the room and the size of the fireplace opening.
2. The woodwork, tiles, and trim should be consistent with other decorative motifs in the room and on the house.
3. The size of the fireplace opening should *never* be changed just for aesthetic purposes. The diameter of the flue and the size of the damper within the chimney relate to the opening dimensions of the fireplace. Altering one of these elements will not only be a violation of code, it will cause the fireplace to function improperly.

If an existing fireplace was thoughtlessly painted, remove the paint from tile, marble, stone, and wood by following the stripping techniques described in Chapter 11. If clinker brick has been painted, the paint-removal process is difficult. The only solution is to sandblast off the paint. This is not only expensive but horribly messy. The equipment costs around $125 a day to rent, and when blasting off the paint you risk marring the surrounding walls with craters and pimples of sand that seem to get stuck everywhere. The process is also dangerous; sand caught in the eyes can cause at least a medical bill, and at most, blindness. When sandblasting, be sure to protect yourself from the flying particles, and by all means keep children and animals out of the area.

Never paint natural brick; to do so is an irreversible action. To remove soot stains from brick, tile, or stone facing, use detergent and water. For stubborn stains, a solution of water and muriatic acid, half and half, should do—definitely wear gloves and goggles.

[POOR DRAW]

A fireplace in poor condition causes two very serious hazards. Insufficient draw permits smoke to fill the room and deoxygenate the breathing air. Poor draw is the result of a blocked or sooty flue, a faulty damper, or an ineffectual ratio of fireplace opening size to flue size. Make sure the damper is operative first before opening it.

REPAIR

You can correct draw problems by reducing the size of the fireplace opening and improving its ratio to the flue size. Experiment by lining the back, sides, or bottom of the firebox with firebrick, minus the mortar. When you discover the installation that works best, make the configuration permanent with fireclay. Remember that building codes require that a firebox be at least 20 inches deep, with top, bottom, and side walls made of at least 4 inches of firebrick, backed by at least 4 inches of common brick, and fireclay joints ¼ inch thick.

Never place a gas heater in a fireplace or direct its gas vent through the chimney. This is architecturally inappropriate, dangerous and *illegal.* Put the gas heater somewhere else, and use the fireplace as another source of heat.

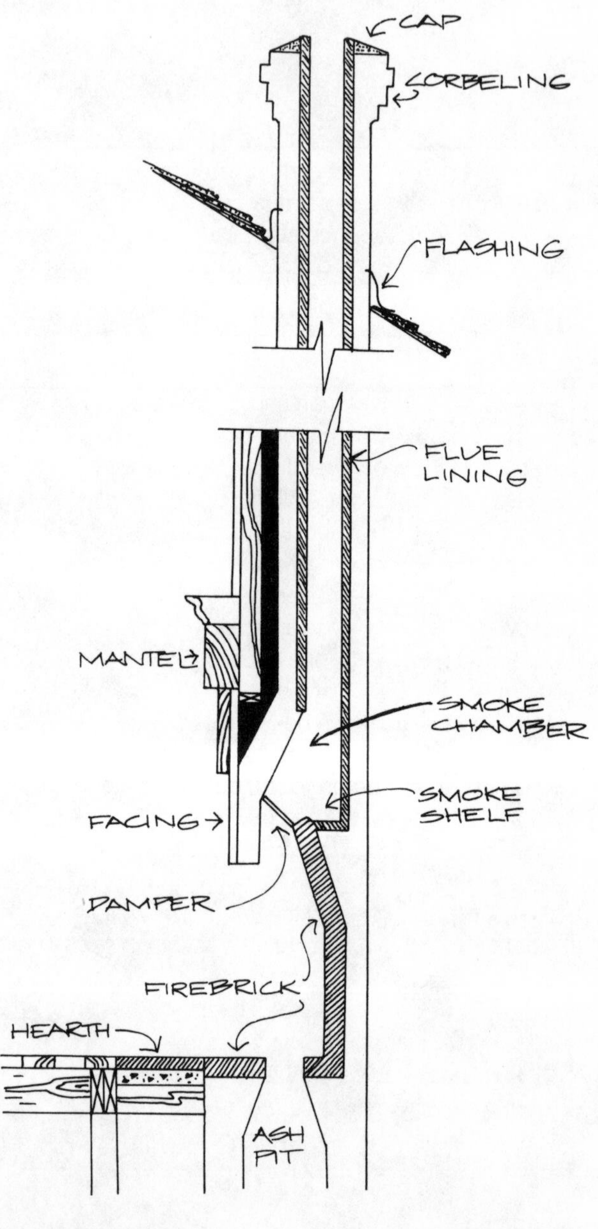

Parts of a fireplace

(LEAKY FLUE)

A leaky flue, another hazard, may allow sparks or flames to contact the building's structural frame and ignite the wood. Leaks result from cracks in the flue liner, crumbling mortar, missing bricks, and the separation of the fireplace from the chimney as a result of settlement. Because accumulated soot is combustible, a dirty chimney sets the stage for a flue fire. Here are the three symptoms of flue hazards and possible repair:

1. *Separation of hearth from floor plane.* Dust and other particles can collect in gaps. Any spark or stray ember that becomes lodged in an unnoticed opening can ignite in no time the surrounding floor, and the fire can spread to the joists below. To set the hearth straight, from below you may have to readjust the slab upon which it sits. (Follow the directions in Chapter 3 for jacking a house to replace or resecure a post.) After you have reset the slab in its proper horizontal position and readjusted its supporting framework, mortar in the gaps so any stray debris will not get caught.

2. *Separation of mantel from wall, as little as ⅛ inch.* The same dangers as with a hearth separation from the floor hold here, only the repair is much simpler. Reset the mantel so it abuts flush with the wall, or install a firmly fitting molding strip along the gap so nothing can make its way into the crevice.

3. *No daylight visible up through the chimney.* (Use a mirror to reflect the rays.) Sometimes bricks loosen and fall inside the chimney shaft and clog the necessary air flow. This can be a bear to correct. Poke a long stick up the chimney to see (or feel) if you can locate the bricks; if you can, your clog is in reaching distance. Keep jabbing until the brick or other blockage comes loose and begins to topple; but watch out for your hands and fingers! If you cannot locate the problem from inside your fireplace, get up on the roof, and use a flashlight to spot the clog. Suspend a weight on a rope or chain, and use it like a vertically working wrecking ball. Again, be sure hands and fingers are clear from below so that tumbling debris does not attack anyone. Once the clog has been removed, the air flow should be sufficient, but the dislodged bricks will have to be compensated for. This is not an easy job because of the sheer awkwardness of trying to work something into a tight spot. Never mind trying to replace the bricks—you may drop them in the process and have to start your unclogging technique all over. Pack a generous amount of cement on the flat blade of a long-handled garden tool like a hoe. Suspend the tool down the shaft. Have a partner push the cement off the blade and into the brick cavity with a long, flat board. Packing it in tightly and smoothly may take several tries, but it is necessary to repair these small problems before the rest of the bricks begin to loosen and upset the stability of your chimney.

As stated, never enlarge a small fireplace opening if the flue size is to remain the same, because the ratio of opening size is critical for maintaining proper draw. The rule of thumb is that the fireplace opening should be not more than 10 to 12 times the area of the flue opening. A short and simple axiom to remember is: width greater than height, and depth of at least 2 feet for good draw.

Do not be frustrated by the size of a small Victorian fireplace. You can use it for burning wood as long as the logs are chopped to fit and burned in reasonably small quantities. (These fireplaces were originally designed for burning coal.)

(STRUCTURAL DAMAGE AND REPAIR)

Repairs will be necessary if the chimney itself is not structurally sound, that is, if the bricks are crumbling, mortar is falling out, or a gap is forming between the shaft and the house. Do not worry too much if the chimney shaft and the house are not tightly joined, because the chimney is a separate structure and undoubtedly will settle independently from the house. It is when the chimney is leaning extensively that you need be concerned. If this is the case, the structure must be rebuilt; consult a masonry contractor, because several codes and much carefuly engineered construction must be met and mastered. This is too great a task for the amateur.

However, replacing or resetting a few bricks is not a difficult feat. (Refer to the section in Chapter 3 regarding the repointing of brickwork and apply the same technique.)

(CLEANING THE CHIMNEY)

To clean a chimney, hire a chimneysweep. Chimneys need cleaning every 15 to 20 years to optimize draw and minimize the danger of a flue fire, but most chimneys have never been cleaned at all. It is well worth the $45 to $65 for the job, considering that a clean chimney will enable you to sleep a little better at night, the time when most chimney fires start.

APPENDIX

(SUPPLIERS)

The following list of suppliers is by no means complete—these are the ones the authors know about. In addition, you will find suppliers of Victorian materials and related items in the yellow pages of your phone book. Many mail order suppliers furnish catalogs; some charge a fee, others do not. It is prudent to write first.

AA-Abbingdon Ceiling Co., Inc.
Tin Ceilings
2149 Utica Ave.
Brooklyn, NY 11234

Acquisition and Restoration Corp.
1226 Broadway
Indianapolis, IN 46202

Alcon Lightcraft Co.
1424 W. Alabama St.
Houston, TX 77006

American Building Restoration
9720 S. 60th St.
Franklin, WI 53132

American Woodcarving
282 San Jose Ave.
San Jose, CA 95125

Arch, Design & Demolishing Co.
Glass, shutters, plaster, etc.
2601 Chartres
New Orleans, LA 70117

Architectural Emphasis, Inc.
1750 Montgomery St.
San Francisco, CA 94111

Architectural Ornaments
P.O. Box 115
Little Neck, NY 11363

Architectural Paneling, Inc.
979 Third Ave.
New York, NY 10022

Architectural Salvage of Santa Barbara
726 Anacapa St.
Santa Barbara, CA 93101

Architectural Specialties, Inc.
850 S. Van Ness Ave.
San Francisco, CA 94109

Arda Inc.
Miscellaneous
77 Church Lane
Philadelphia, PA 19144

William Armstrong & Associates
Craftsmen in Stained Glass
20 Dalton St.
Newburyport, MA 01950

Art Directions
Doors, mantels, columns
6120 Delmar Blvd.
St. Louis, MO 63112

Artifacts, Inc.
702 Mt. Vernon Ave.
Alexandria, VA 22301

The Astrup Company
2937 W. 25th St.
Cleveland, OH 44113

Authentic Designs
330 E. 75th St.
New York, NY 10021

Authentic Reproduction Lighting Co.
P.O. Box 218
Avon, CT 06001

Ball and Ball
Hardwood, lighting
463 W. Lincoln Highway
Exton, PA 19341
(catalog $4)

A. W. Baker Restorations, Inc.
670 Drift Road
Westport, MA 02790

Bare Wood Inc.
Brownstone hardware, mantels, fretwork, etc.
141 Atlantic Ave.
Brooklyn, NY 11201

Barney Brainum-Shanker Steel Co.
Metal ceilings and wall coverings
70-32 83rd St.
Glendale, NY 11227

Charles Bellinger/Architectural Components
P.O. Box 246
Leverett, MA 01054

Bendix Mouldings, Inc.
235 Pegasus Ave.
Northvale, NY 07647

Bennington Bronze
147 S. Main St.
White River Junction, VT 05001

Bernardini Iron Works, Inc.
418 Bryant Ave.
Bronx, NY 10474

Beveled Glass Industries
979 Third Ave.
New York, NY 10022

Black Millwork Co., Inc.
Lake Avenue
Midland Park, NJ 07432

Blaine Window Hardware, Inc.
1919 Blaine Dr.
Hagerstown, MD 21740

By Gone Era
Miscellaneous
4783 Peachtree Rd.
Atlanta, GA 30341

Capability Brown Limited
130 W. 28th St.
New York, NY 10001

Cape Cod Cupola Co., Inc.
78 State Rd.
North Dartmouth, MA 02747

Ceilings, Walls & More, Inc.
Box 494, 124 Walnut St.
Jefferson, TX 75657

Celestial Roofing
1710 Thousand Oaks Blvd.
Berkeley, CA 94702

Colonial Restoration Materials
Route 123
Stoddard, NH 30464

Colonial Tin Craft
7805 Railroad Ave.
Cincinnati, OH 45243

Colonial Wood
16 Water St.
Clinton, NJ 08809

Cumberland Wood Craft Co., Inc.
Victorian millwork
R.D. 5 Box 452
Carlisle, PA 17013
(catalog $2)

Daly's Wood Finishing Products
Woodwork refinishing
1121 N. 36th
Seattle, WA 98103

Dana-Deck Inc.
Exterior shingles
P.O. Box 78
Orcas, WA 98280

Decorators Supply Corp.
3610-125 Morgan St.
Chicago, IL 60609

Early New England Restorations Joinery and Forge Works
P.O. Box 45
Mansfield Depot, CT 06251

18th Century Hardware Co.
131 East 3rd St.
Derry, PA 15627

Fife's Woodworking & Mfg. Co.
Main Street
Northwood, NH 03261

Fine Metalwork
Deanne F. Nelson
Box 43
Bishop Hill, IL 61419

Grammar of Ornament
Painters, stencils, decorative plasterwork
2122 W. 28th Ave.
Denver, CO 80211

Gorsuch Foundry Co., Inc.
Hardware
120 E. Market St.
Jeffersonville, IN 47130

Guilfoy Cornice Works
1234 Howard St.
San Francisco, CA 94005

Hallelujah Redwood Products
Wooden ornamentation
39500-J Comptche Rd.
Mendocino, CA 95460
(catalog $1)

The House Carpenters
Box 217
Shutesbury, MA 01072

Steve Kayne Hand Forged Hardware
17 Harmon Place
Smithtown, NY 11787

KB Moulding, Inc.
508A Lakefield Rd.
East Northport, NY 11731

Lead Glass Co.
14924 Beloit Snodes Rd.
Beloit, OH 44609

Locks & Handles
8 Exhibition Rd.
London SW7 2HF
England

Lone Star Door & Millwork
Victorian doors
P.O. Box 607
Irving, TX 75060

Materials Unlimited
4100 Morgan Rd.
Ypsilanti, MI 48197

New York Flooring
340 E. 90th St.
New York, NY 10028

Northeast American Heritage Co.
77 Washington St., N., Suite 502
Boston, MA 02114

Craig Nutt Fire Wood Works
2308 Sixth St.
Tuscaloosa, AL 35401

Old Carolina Brick Co.
Rt 9, Box 77
Majolica Road
Salisbury, NC 28144

Old House Supplies
Pandora's Antiques
2014 Old Philadelphia Pike
Lancaster, PA 17602

Old Mansions Co.
1305 Blue Hill Ave.
Mattapan, MA 02126

Olde New England Masonry
27 Hewitt Rd.
Mystic, CT 06355

Old Town Restorations
158 Farrington St.
St. Paul, MN 55102

Old Wood Moulding & Finishing Co., Inc.
115 Allen Blvd.
Farmingdale, NY 11735

Preservation Associates, Inc.
P.O. Box 202
Sharpsburg, MD 21782

Preservation Resource Center
Lake Shore Rd.
Essex, NY 12936

Preservation Resource Group
5619 Southampton Dr.
Springfield, VA 22151

Rainbow Art Glass Co.
49 Shark River Rd.
Neptune, NH 07753

Rejuvenation House Parts Co.
4343 N. Albina Ave.
Portland, OR 97217

The Renovation Source, Inc.
3512-14 N. Southport Ave.
Chicago, IL 60657

The Renovators Co.
Box 284
Patterson, NY 12563

Renovator's Supply
71 Northfield Rd.
Millers Falls, MA 01349

Restoration A Speciality
6127 N.E. Rodney St.
Portland, OR 97211

Restorations, Ltd.
Jamestown, RI 02835

Restorations Unlimited, Inc.
24 W. Main St.
Elizabethville, PA 17023

San Francisco Victoriana
Moldings, ceilings (plaster)
2245 Palou Ave.
San Francisco, CA 94124

Silverton Victorian Millworks
Millwork
Box 523
Silverton, CO 81433
(catalog $2)

Standard Trimming Corp.
1114 First St.
New York, NY 10021

Stencil Art
232 Amazon Place
Columbus, OH 43214

Stencil Specialty Co.
377 Ocean Ave.
Jersey City, NH 07305

United House Wrecking Corp.
328 Selleck St.
Stamford, CT 06902

Victorian Building & Repair
Victorian restoration
RR 1B, Box 162-A
Compton, IL 61318

Victorian D'Light
Lighting
535 W. Windsor Rd.
Glendale, CA 81204

Victorian Reproduction Ent. Inc.
Lighting fixtures, hardware, millwork, plumbing
Dept. 1A
1601 Park Avenue South
Minneapolis, MN 55404

Vintage Wood Works
Victorian ornamentation
Rt 2 Box 68R
Quinlan, TX 75474
(catalog $1)

Woodcraft Supply Corp.
313 Montvale Ave.
Woburn, MA 01801

The Wrecking Bar
292 Moreland Ave. N.E.
Atlanta, GA 30307

The Wrecking Bar
2601 McKinney Ave.
Dallas, TX 75204

Yankee Craftsman
357 Commonwealth Road
Wayland, MA 01778

Ye Olde Mantel Shoppe
3800 NE 2nd Ave.
Miami, FL 33137

[PERIODICALS]

American Preservation (bimonthly)
P.O. Box 589
Martinsville, NJ 08836

Americana (bimonthly)
Americana Subscription Office
381 W. Center St.
Marion, OH 43302

Bulletin (monthly) and *19th Century* (bimonthly)
The Victorian Society in America
East Washington Square
Philadelphia, PA 19106

Colonial Homes (bimonthly)
P.O. Box 10159
Des Moines, IA 50350

Home Restoration (bimonthly)
P.O. Box 327
Gettysburg, PA 17325

The Magazine Antiques
551 Fifth Avenue
New York, NY 10017

Old-House Journal (monthly)
69A Seventh Ave.
Brooklyn, NY 11217

Preservation News (monthly) and *Historic Preservation* (bimonthly)
The National Trust for Historic Preservation
1785 Massachusetts Ave., N.W.
Washington, DC 20036

INDEX

References to illustrations are in *italic* type.